Learning and Analytics in Intelligent Systems

Volume 19

Series Editors

George A. Tsihrintzis, University of Piraeus, Piraeus, Greece

Maria Virvou, University of Piraeus, Piraeus, Greece

Lakhmi C. Jain, Faculty of Engineering and Information Technology, Centre for Artificial Intelligence, University of Technology, Sydney, NSW, Australia; KES International, Shoreham-by-Sea, UK; Liverpool Hope University, Liverpool, UK

The main aim of the series is to make available a publication of books in hard copy form and soft copy form on all aspects of learning, analytics and advanced intelligent systems and related technologies. The mentioned disciplines are strongly related and complement one another significantly. Thus, the series encourages cross-fertilization highlighting research and knowledge of common interest. The series allows a unified/integrated approach to themes and topics in these scientific disciplines which will result in significant cross-fertilization and research dissemination. To maximize dissemination of research results and knowledge in these disciplines, the series publishes edited books, monographs, handbooks, textbooks and conference proceedings.

More information about this series at http://www.springer.com/series/16172

Maria Virvou · Hiroyuki Nakagawa ·
Lakhmi C. Jain
Editors

Knowledge-Based Software Engineering: 2020

Proceedings of the 13th International
Joint Conference on Knowledge-Based
Software Engineering (JCKBSE 2020),
Larnaca, Cyprus, August 24–26, 2020

 Springer

Editors
Maria Virvou
Department of Informatics
University of Piraeus
Piraeus, Greece

Hiroyuki Nakagawa
Graduate School of Information Science
and Technology
Osaka University
Osaka, Japan

Lakhmi C. Jain
Faculty of Engineering and Information
University of Technology Sydney
Stonyfell, SA, Australia

ISSN 2662-3447   ISSN 2662-3455   (electronic)
Learning and Analytics in Intelligent Systems
ISBN 978-3-030-53951-1   ISBN 978-3-030-53949-8   (eBook)
https://doi.org/10.1007/978-3-030-53949-8

This Springer imprint is published by the registered company Springer Nature Switzerland AG
The registered company address is: Gewerbestrasse 11, 6330 Cham, Switzerland

Preface

This volume summarizes the works and new research results presented at the **13th International Joint Conference on Knowledge-based Software Engineering (JCKBSE 2020)**, which took place in August 24–26, 2020. JCKBSE2020 was originally scheduled to take place in Larnaca, Cyprus. Unfortunately, the COVID-19 pandemic forced us to reschedule JCKBSE2020 as an online conference.

JCKBSE is a well-established international biennial conference that focuses on the applications of artificial intelligence on software engineering. The 13th International Joint Conference on Knowledge-based Software Engineering (JCKBSE 2020) was organized by Hiroyuki Nakagawa, Graduate School of Information Science and Technology, Osaka University, Japan; and George A. Tsihrintzis and Maria Virvou, Department of Informatics, University of Piraeus, Greece.

This year, pretty much like every year, the majority of submissions originated from Japan, while Greece was second. The submitted papers were rigorously reviewed. Finally, 15 papers were accepted for presentation at the JCKBSE2020 and inclusion in its proceedings. The papers accepted for presentation in JCKBSE2020 address topics such as the following:

- Architecture of knowledge-based systems, intelligent agents and softbots
- Architectures for knowledge-based shells
- Automating software design and synthesis
- Decision support methods for software engineering
- Development of multi-modal interfaces
- Development of user models
- Development processes for knowledge-based applications
- Empirical/evaluation studies for knowledge-based applications
- Intelligent user interfaces and human–machine interaction
- Internet-based interactive applications
- Knowledge acquisition
- Knowledge engineering for process management and project management

- Knowledge management for business processes, workflows and enterprise modeling
- Knowledge technologies for semantic web
- Knowledge technologies for service-oriented systems, Internet of services and Internet of things
- Knowledge technologies for web services
- Knowledge-based methods and tools for software engineering education
- Knowledge-based methods and tools for testing, verification and validation, maintenance and evolution
- Knowledge-based methods for software metrics
- Knowledge-based requirements engineering, domain analysis and modeling
- Methodology and tools for knowledge discovery and data mining
- Ontologies and patterns in UML modeling
- Ontology engineering
- Program understanding, programming knowledge, modeling programs and programmers
- Software engineering methods for Intelligent Tutoring Systems
- Software life cycle of intelligent interactive systems
- Software tools assisting the development of intelligent systems

In addition to technical paper presenters, in JCKBSE2020, we had the following distinguished researchers as keynote speakers:

1. Prof.-Dr. Nikolaos Bourbakis, Wright State University, USA
2. Prof.-Dr. Haruhiko Kaiya, Kanagawa University, Japan

We would like to thank the authors for choosing JCKBSE2020 as the forum for presenting the results of their research. Additionally, we would like to thank the reviewers for taking the time to review the submitted papers rigorously. For putting together the website of JCKBSE2020 and for managing the conference administration system and coordinating JCKBSE2020, we would like to thank **Easy Conferences Ltd., Nicosia, Cyprus.**

We would like to thank **Dr. Thomas Ditzinger** for agreeing to publish the JCKBSE2020 proceedings in the form of a volume in the ***Learning and Analytics in Intelligent Systems*** series of Springer (https://www.springer.com/series/16172), which he edits. Finally, we would like to thank the **Springer personnel** for their wonderful job in producing this volume of proceedings.

The JCKBSE2020 Honorary Chairs

Shuichiro Yamamoto, Nagoya University, Japan
Lakhmi C. Jain, University of Canberra, Australia, and University of Technology Sydney, Australia

The JCKBSE2020 General Chairs

George A. Tsihrintzis, University of Piraeus, Greece
Hiroyuki Nakagawa, Osaka University, Japan

The JCKBSE2020 Program Chairs

Maria Virvou, University of Piraeus, Greece
Hiroyuki Nakagawa, Osaka University, Japan

Keynote Speech

A SPN Holistic Approach for Automatic Deep Understanding of Technical Documents

Nikolaos G. Bourbakis

Wright State University, Ohio, USA
Nikolaos.bourbakis@wright.edu

Abstract. For many decades, scientists and practitioners have been developing methodologies for solving simple or complex scientific problems. Lately, they have been creating synergies of different methodologies in order to address and solve more complex problems, but they have to face issues of compatibility among the different modalities that compose these scientific problems. One of these complex problems is the deep processing and understanding of technical documents, which are composed by different modalities, such as natural language text, system diagrams, mathematical formulae, algorithms, tables, graphics, pictures. Humans have the ability to understand, process and associate these modalities to significantly contribute to the overall deeper understanding of the technical document and the knowledge represented in it. Thus, this talk offers a possible solution to these multi-modal complex problems by expressing all the modalities into the same medium (in this case the Stochastic Petri-nets (SPN) model) for a synergistic and automatic deep understanding of technical documents (TDs). Specifically, all the modalities are expressed into Natural Language Sentences and into SPN models for better associations and deeper understanding.

Biography

Dr. Nikolaos Bourbakis (IEEE Life Fellow) is currently a Distinguished Professor of Information & Technology and the Director of the Center of Assistive Research Technologies (CART) at Wright State University, Ohio, USA. He is the Founder and the EIC of three International Journals (*Artificial Intelligence Tools (World Scientific Publ.), Monitoring and Surveillance Tech Research (IGI-Global, Publ.) and Bioengineering & Bioinformatics (EAI Transactions)*) and the Founder and Steering Committee Chair of several International IEEE Computer Society Conferences (ICTAI, BIBE, IISA), Symposia and

Workshops. He pursues research in Assistive Technologies, Applied Artificial Intelligence, Bioengineering, Information Security and Distributed Processing funded by USA and European agencies. He has published extensively in IEEE and International Journals and Conferences, and he has supervised several dozens of PhD students who have successfully graduated. His research work has been internationally recognized, and he has won several prestigious awards in IEEE, Universities and Societies.

Quantitative Analysis of Goal-Oriented Requirements Models

Haruhiko Kaiya

Kanagawa University, Hiratsuka 259-1293, Japan
kaiya@acm.org

Abstract. An information system is developed and embedded into a dairy activity to satisfy requirements of people in the activity. Because the system is expected to improve the activity, we have to predict the extent of such improvement. Goal-oriented requirements models are useful to represent the relationships among systems and people in a business or life activity. We have proposed extended goal-oriented requirements models to predict how well the system improves the activity of the people. In this talk, we briefly introduce typical goal-oriented requirements models and their extensions suitable for quantitative analysis. Finally, we show several issues of future challenges.

Biography

Haruhiko Kaiya is a Professor in the Faculty of Science at Kanagawa University, Japan. He is interested mainly in requirements engineering, especially goal-oriented models and ontologies. He was a member of program committee in International Requirements Engineering Conferences (RE) and International Conferences on Advanced Information Systems Engineering (CAiSE).

Contents

Quantitative Analysis of Goal Oriented Requirements Models

Haruhiko Kaiya[(⊠)]

Kanagawa University, Hiratsuka 259-1293, Japan
kaiya@acm.org

Abstract. An information system is developed and embedded into a dairy activity to satisfy requirements of people in the activity. Because the system is expected to improve the activity, we have to predict the extent of such improvement. Goal oriented requirements models are useful to represent the relationships among systems and people in a business or life activity. We have been proposed extended goal oriented requirements models to predict how well the system improves the activity of the people. In this talk, we briefly introduce typical goal oriented requirements models, and their extensions suitable for quantitative analysis. Finally, we show several issues of future challenges.

Keywords: Goal Oriented Requirements Analysis · Quality requirements · Metrics

1 Introduction

We are interested in developing good information systems and services. Traditionally, the voice of stakeholders should be primary taken into account to develop them. Recently, many kinds of people and systems are connected and related with each other. Many different activities are then performed simultaneously. Therefore, we can regard the world or its part itself as a system, i.e. a Socio-Technical Systems (STS). It is not easy to find all or main stakeholders of the STS in advance. Therefore, the voice of the stakeholders cannot be directly taken into account. Instead, we can predict the extent and the direction of changes caused by systems to be introduced. This prediction will make the potential stakeholders aware of their gain and loss, and we can then obtain their voice to fit the system into the stakeholders.

Goal oriented requirements models are useful tools to predict such issues because the stakeholders can directly know whether their goals are satisfied or not. To apply quantitative analysis in the goal oriented requirements models, stakeholders can know the extent of changes related to their goals caused by the system. On the basis of the rationale above, we have studied quantitative analysis of goal oriented requirements models. We introduce its results in this talk.

M. Virvou et al. (Eds.): JCKBSE 2020, LAIS 19, pp. 1–6, 2020.
https://doi.org/10.1007/978-3-030-53949-8_1

The rest of this note is organized as follows. In the next section, we briefly introduce famous and general notations of goal oriented requirements models. Our models are based on them. We then introduce our extensions of goal oriented requirements models in Sect. 3. We summarize the future challenges mainly related to quantitative analysis of requirements. Finally, we remark that the value and usefulness is quite important in engineering.

2   Typical Goal-Oriented Requirements Models

Here we briefly explain two famous goal oriented models KAOS [6] and Istar (usually written as i*) [11] although other models and concepts such as GSN [10] and GQM [9] exist. Both KAOS and Istar have a lot of characteristics respectively, but we only focus on its core idea in this section.

KAOS contains several different models for requirements. In addition, formal logics can be used to specify elements in the models. Its core model is a goal model. A goal model in KAOS is very usual such that a top goal is iteratively decomposed into several sub-goals. The final goal model is then represented in a tree or a directed asynchronous graph of goals. We think the core idea of KAOS is the responsibility assignment to the leaf goals. When we develop a goal tree, there is no clear criteria of terminating goal decomposition before KAOS. However in KAOS, how to find the top goal is almost out of scope. Another core idea of KAOS is the definition of what a goal is. In several goal oriented models including ours [3,7], the definition of a goal is vague. In a worst case, a goal is represented in a function or an action such as "doing something". In KAOS, the states of the world are focused to define a goal. By using such states, KAOS then defines several types of goals such as maintaining a state, reaching a state and avoiding the transitions to a state.

Istar is another famous goal oriented model. In Istar, actors such as human, organizations and machines including information systems are initially focused. A goal is then defined a dependency relationship between an actor and another, each of which are called "depender" and "dependee". A depender is an actor who wants to achieve a goal, and a dependee is an actor who will achieve the goal. This dependency relationship can clearly specify the source of initial goals although the source is not clear in KAOS. This dependency relationship also represents the delegation of a goal to another actor. Because each actor has his/her/its own rationale, a delegated goal to a dependee is usually decomposed into several goals in the dependee and some of them are recursively delegated to other actors. If a goal delegated to an actor can be achieved by the actor, the actor is regarded to have the means of achieving the goal. The delegated goal is thus neither decomposed nor delegated anymore. This mechanism is similar to the responsibility in KAOS. Although Istar contains contribution relationships among non-functional and/or quality goals, the characteristics is not so outstanding.

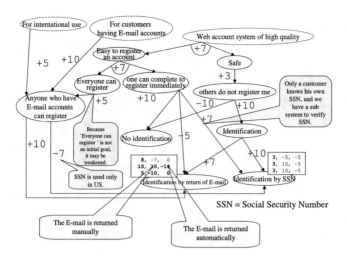

Fig. 1. An example of AGORA model [3]

3   Quantitative Analysis of Goal Models

Attributed Goal Oriented Requirements Analysis (AGORA) [3,7] is our first attempt to introduce quantitative analysis in goal oriented requirements models. AGORA uses KAOS-style goal models as shown in Fig. 1. In AGORA, following two types of values are introduced into goal models.

- Contribution value: The value specifies the degree of the sub-goal's contribution to its parent goal.
- Preference matrix: The matrix specifies the discordance among stakeholders about evaluating a goal.

Using the contribution values, the quality of the goal model is measured from six viewpoints, i.e. correctness, unambiguity, completeness, inconsistency, modifiability and traceability. A preference matrix of a goal lets us know whether the interpretation of the goal is consistent among the stakeholders or not.

More than 10 years later, we proposed Goal Dependency Model with Attributes (GDMA) [2,5]. GDMA is a variation of Istar as shown in Fig. 2. In GDMA, soft-goals and contribution links in Istar are not used. Instead, we add the quality characteristics such as "quickly" or "accurately" to each goal because we understood quality attributes should be always qualified some functional requirements. We then introduce the following two values for each quality characteristic in each goal between a depender and a dependee.

- want-level: This value is attached to the link between a depender and a goal. When the goal has more than two quality characteristics, the vector of want-levels is attached. This value shows the extent how much the corresponding quality characteristic is wanted by the depender.

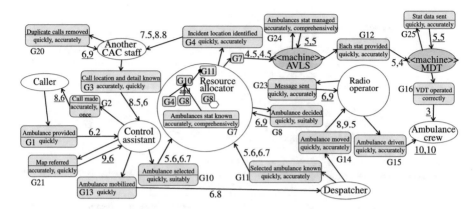

Fig. 2. An example of GDMA model about the London Ambulance Service (each want-level is omitted because it is always 10 in this case.) [5]

- can-level: This value is attached to the link between a goal and a dependee in the same way as the want-level. This value shows the extend how much the corresponding quality characteristic can be achieved by the dependee.

Each level takes 1 to 10. The value 1 means "not so wanted" or "cannot achieve". The value 10 means "extremely wanted" or "completely achieve". For example in Fig. 2, we can find a goal dependency between AVLS and MDT at the top right. This dependency represents AVLS wants MDT to provide the status of each ambulance quickly and accurately. Although AVLS wants MDT to do extremely quickly and accurately, MDT does it not so quickly (level 5) and a little bit inaccurately (level 4).

By using the topology of the model and levels above, we can derive the following four metrics.

- Average Number of Wants (ANW): This metric represents the number of human goals achieved in this STS. Note that derived goals and machines goals are not counted.
- Average Number of Cans (ANC): This metric represents the number of goals, which should be achieved by human.
- Average Gains of Wants (AGW): This metric represents how well human goals are achieved in average.
- Average Gains of Cans (AGC): This metric represents how much human has to achieve goals.

These metrics can be calculated in each version of a STS model. We decide a new version is better than ever if ANW and AGW increase, and ANC and AGC decrease in the new version. This decision is based on the assumption that human wants others to achieve his/her goals more and better than ever, but he/she does not want to contribute to others.

4   Future Challenges

Our goal oriented requirements models contribute to analyzing a STS quantitatively. However, it still remains problems to be solved.

1. How to decide numeric attributes?
   It is not easy to decide numerical values in AGORA and GDMA. Currently, these values should be decided subjectively. We expect the history of value decision contributes to the next decision.
2. How to decide the boundary of a STS?
   Even an identified STS becomes better than ever with some information systems, we cannot guarantee the systems never badly affect the outside the STS. We have to carefully decide the boundary of a STS, or to take mutual effects among several STSs. We are now investigating the interactions among several different systems [4].
3. What is the ultimate goals of stakeholders?
   To identify the ultimate goals of human, we can introduce suitable machines into some STS. On the basis of a famous scientific book [1], we suppose the ultimate goal of human is to preserve their genes. Most social and economic activities and goals seem to be just means to achieve this ultimate goal. This is one of the reasons why the assumption of metrics evaluation in GDMA is selfish. We want to validate goals delegated to machines on the basis of such ultimate goals.
4. How to introduce the models into the real world problems?
   In many software and requirements engineering researches, it is very hard to bring some method or a technique into practice. We of course want to bring our goal-oriented requirements models into practice.

5   Final Remarks

According to a famous dictionary in Japan [8], "engineering" is a general term of applied science and technology to improve the productivity of *the manufacturing industries*. The knowledge of fundamental sciences is used for such improvement. By its definition, "manufacturing industry" has to transform the materials and/or the parts into valuable and/or useful things. Therefore, "valuable" and "useful" are the core concepts in any kinds of engineering although they are quite subjective and depend on the whim of people. Software products and services are the same as the traditional manufacturing products in the sense that the value and the usefulness are important. In knowledge engineering, software engineering, requirements engineering and information system engineering, we think we always have to take value and usefulness into account. Goal oriented requirements models help us to remember such a fundamental issue. Especially, our extensions directly contribute to measuring values and usefulness.

References

1. Dawkins, R.: The Selfish Gene, 40th anniversary edn. Oxford University Press, Oxford (2016)
2. Kaiya, H., Haga, K.: A CASE tool for goal dependency model with attributes based on an existing UML editor. In: Knowledge-Based and Intelligent Information & Engineering Systems: Proceedings of the 21st International Conference KES-2017, Marseille, France, September 6–8, 2017, pp. 1196–1205 (2017). https://doi.org/10.1016/j.procs.2017.08.033
3. Kaiya, H., Horai, H., Saeki, M.: AGORA: attributed goal-oriented requirements analysis method. In: 10th Anniversary IEEE Joint International Conference on Requirements Engineering (RE 2002), Essen, Germany, September 9–13, 2002, pp. 13–22 (2002). https://doi.org/10.1109/ICRE.2002.1048501
4. Kaiya, H., Muto, R., Nagano, K., Yoshida, M.: Mutual requirements evolution by combining different information systems. In: Proceedings of the 23rd Pan-Hellenic Conference on Informatics, PCI 2019, Nicosia, Cyprus, November 28–30, 2019, pp. 159–162 (2019). https://doi.org/10.1145/3368640.3368662
5. Kaiya, H., Ogata, S., Hayashi, S., Saeki, M.: Early requirements analysis for a socio-technical system based on goal dependencies. In: New Trends in Software Methodologies, Tools and Techniques - Proceedings of the Fifteenth SoMeT_16, Larnaca, Cyprus, September 12–14, 2016, pp. 125–138 (2016). https://doi.org/10.3233/978-1-61499-674-3-125
6. van Lamsweerde, A.: Requirements Engineering: From System Goals to UML Models to Software Specifications. Wiley, New York (2009)
7. Saeki, M., Hayashi, S., Kaiya, H.: A tool for attributed goal-oriented requirements analysis. In: 24th IEEE/ACM International Conference on Automated Software Engineering, ASE 2009, Auckland, New Zealand, November 16–20, 2009, pp. 674–676 (2009). https://doi.org/10.1109/ASE.2009.34
8. Shinmura, I. (ed.): Kojien. Iwanami Shoten, 7th edn. (2018), Japanese Dictionary, in Japanese
9. Solingen, R.V., Berghout, E.: The Goal/Question/Metric Method: A Practical Guide For Quality Improvement Of Software Development. McGraw-Hill, New York (1999)
10. Spriggs, J.: GSN - The Goal Structuring Notation: A Structured Approach to Presenting Arguments. Springer, London (2012)
11. Yu, E., Giorgini, P., Maiden, N., Mylopoulos, J.: Social Modeling for Requirements Engineering. The MIT Press, Cambridge (2010)

Revisiting Principles and Challenges
in Natural Language Programming

Juliano Efson Sales[1,2(✉)], André Freitas[3], Douglas Oliveira[4],
Adamantios Koumpis[5], and Siegfried Handschuh[2]

[1] Department of Computer Science and Mathematics, University of Passau,
Innstraße 43, 94032 Passau, Germany
[2] Institute of Computer Science, University of St. Gallen,
Rosenbergstrasse 30, 9000 St. Gallen, Switzerland
juliano.sales@unisg.ch
[3] School of Computer Science, University of Manchester, Kilburn Building,
Oxford Road, Manchester M13 9PL, UK
[4] Architecture and Analytics Group, VLI Logística,
Rua dos Tapuias 49, Belo Horizonte, Brazil
[5] Institute of Digital Enabling, Bern University of Applied Sciences,
Brückenstrasse 73, 3005 Bern, Switzerland

Abstract. Automation has faced the risk of reducing its pace due to
the shortage of information technology professionals. Although part of
the programming demand can be addressed by simple compositions of
high-level functions and data flows, non-technical professionals are still
unable to build their own software given the intrinsic complexity of coding. Among other types of end-user development, natural language programming has emerged as a strong candidate to fill this gap by developing
methods and tools to allow end users to program. The paper revisits some
principles of evaluation of traditional programming languages and analyses the new challenges to deliver an effective end-user development platform based on aspects of natural language processing, human-computer
interaction, software engineering, and programming education. We advocate that an effective end-user platform is essentially hybrid, combining
features from different branches of the end-user development research,
having, however, a search mechanism with semantic capabilities at its
centre.

Keywords: Very high-level languages · Natural language interfaces ·
Natural language programming

1 Introduction

The social demand for automation has increased yearly pushed by the advances
in information technologies (IT) and the decreasing cost of hardware, also boost-
ing, as a side effect, the demand for IT professionals. By 2020, Europe can suffer
from a shortage of up to 500,000 IT professionals [14], whereas the United States

© Springer Nature Switzerland AG 2020
M. Virvou et al. (Eds.): JCKBSE 2020, LAIS 19, pp. 7–19, 2020.
https://doi.org/10.1007/978-3-030-53949-8_2

already has today more than 220,000 open positions for software developers without qualified people to fill them [49].

Many programming tasks, however, require neither the exercise of deeper computer science skills nor sophisticated software engineering, consisting of the composition of functions and the semantic harmonisation of the data flows across different components. This is typically the case for analytic tasks or small-scale business processes, which do not have high requirements for robustness and performance under critical conditions. However many non-technical professionals such as domain experts and business analysts are still unable to build their own software due to the intrinsic complexity of coding, e.g. the formalism of traditional programming languages and the technical knowledge to understand and integrate multiple software artefacts [42].

Among other types of end-user development, natural language programming has emerged as a strong candidate to fill this gap by developing methods and tools to allow end-users to program [25]. Considering aspects from the fields of *natural language processing, human-computer interaction, software engineering* and *programming education*, this paper analyses the challenges and opportunities of hybrid end-user development platforms where the natural language plays a central role. To support this analysis, we revisit the principles of evaluation of traditional programming languages, adjusting and extending them to this context.

The rest of this paper is organised as follows. In Sect. 2, the paper starts arguing for putting natural language at the centre of end-user-development debate. To contextualise the further discussion, Sect. 3 brings a brief survey on natural language programming, presenting the early paradigms, the contribution of semantic technologies and the contemporary set of machine learning supported techniques. Section 4 defines both the typical task and the typical user considered in the study. Sections 5 and 6 are the central part of our contributions, describing respectively the principles and challenges that an effective natural language programming model faces to proper support non-technical users to write code. Finally, Sect. 7 summarises the study and highlights the potential and social impacts of the new generation of end-user programming platforms.

2   Natural Language at the Centre

Since its debut, end-user development (EUD) has matured four main branches of research: *visual programming, programming by example, code synthesis* and *natural language programming* (*NL programming* from now), each of them having advances and disadvantages.

Visual programming, in which blocks and graphics provide an easier syntax to end users, allows the developers to start fast, but its effectiveness in the long run is limited due to (*i*) restrictions in its expressivity and (*ii*) the change of the *notional machine* [18]. Notional machine is the abstraction of an idealised computer, from the run-time point of view, aiming at describing the semantics of the programs and defining the typical behaviour of the code instructions

[48]. Furthermore, visual programs are frequently constrained to a sub-set of programming constructors, which limit the type of computation that can be described. To gain expressivity, this scenario can eventually force to a change to a textual language. As there is significant difference between the notional machines of visual programming environments and textual environments, the previously programming experience does not help constructing a gentle leaning curve [23].

Programming by example [17] and *code synthesis* [38] have evolved significantly in recent years allowing respectively the inference of programs from some instances input-output examples and the *"translation"* of natural language into code. These approaches, however, suffer from the lack of *graceful repairing*, i.e. in the case of misbehaviour, they don't provide mechanism to correct or improve the results, except by redoing the task from scratch, without taking advantage of the previous results.

Having a textual language at the centre of the end-user development environment from the beginning, and maintaining it as the main programming interaction method, allows the user to keep the same notional machine during their programming experience, without preventing the use of visual aids and functions inspired by programming-by-example as auxiliary tools.

3   Related Work

Sammet [44] was one of the first voices to advocate that technical languages are an overhead, preventing end-user from focusing on the problem at hand. However, the idea that human-computer languages should be designed from the perspective of a natural language did not find a consensus.

Dijkstra, for instance, called NL programming *"a foolish idea"* [13]. His main plea is supported by the fact that a machine does exactly what it is instructed to do and natural language is essentially vague, imprecise and ambiguous [12]. From his point of view, in addition to demanding the construction of a much more complex *"compiler"*, his main objection relies on the comparison with mathematics, where the lack of proper formalism would stop the development of the natural sciences.

Despite the initial criticism, the idea of using natural language to program was not completely rejected and the area evolved in the following decades.

3.1   The First Attempts

The prototypes developed by Ballart *et al.* [1] and Biermann *et al.* [3] represent the typical initial effort in constructing a NL programming environment. These first attempts targeted a general purpose programming language able to execute mathematical operations such as matrix manipulation.

Related initiatives also aimed at providing natural language interfaces to execute operations generally demanding technical knowledge such as Guida & Tasso's work [16] to integrate a natural language interface to databases, Maas and Suppes' [28] to control an instructable robot, and Manaris *et al.*'s [29] to manage an operating system commands.

Commands such as *"Double x and store in y"*, which was the focus of the first initiatives, demonstrate the reason why part of the scientific community was opposed to NL programming [3]. However, with the growing availability of more complex software, APIs with a higher functional complexity which could better represent the level of abstraction present in the end-user's mental model, and so, making a natural language interface closer to her/his intents.

3.2   On the Play of Semantic Web

The consolidation of the Semantic Web [2] established an important milestone towards the semantic description of web services, where ontologies became a key component of architectures for natural language understanding systems.

Englmeier *et al.* [15] propose an approach to process storybooks written in natural language to interpret the commands, and translating them to a choreography script. Relying on the similarity between an ontology-based service description and a formal representation of the user request, Bosca *et al.* [7] designed a method in which a restricted set of words and sentence templates compose services. Lim *et al.* [26] also present a similar work making use of keywords and main verbs to map a natural language command to an OWL-S ontology. In addition to the ontology, the works of Pop *et al.* [36] and Sangers *et al.* [45] also make use of linguistic resources by respectively enriching the vocabulary with WordNet and applying WordNet-similarities metrics [31].

The work of Wong [52] allows end-user developers to collect information from different websites using their web services APIs. The tool generates a sort of mash-up, where complex tasks can be divided into simpler ones that are processed using data flow metaphors.

Aiming at better understanding the requirements of end users and attempting to simplify service composition, Namoun *et al.* [32] propose several guidelines for the development of service decomposition tools.

Unanimously, the main contribution of these works is the design of a software architecture able to compose distributed services using a natural language interface, even though, the semantic methods present rudimentary strategies to comprehend the natural language utterances, grounding mostly in a controlled vocabulary, keywords and POS-tags rules. On the other hand, these works highlight the concept of the choreography of high-level components and consolidate the notion of a distributed architecture to create web-scale applications.

3.3   The Representation Learning Perspective

Machine learning models together with annotated data [19] and linguistics resources [31] established new levels of maturity for tasks such as *semantic parsing, question answering* and *text entailment*, whose principles are largely applied in *NL programming platforms* [30].

New classes of machine learning models, namely, recurrent neural networks such as *Long Short-Term Memory* (LSTM) and *Gated Recurrent Unit* (GRU), and attention-based models like Transformers [51], have been attracting attention over the years due to their performance on natural language problems, producing robust representations of language models, especially when aligned with word embeddings [9,43].

Supported by these technologies, big players such as *Google*, *Amazon* and *Apple* have built assistants that trigger atomic requests and answer simple questions by voice [27]. For instance, the Amazon team defined a language for the virtual assistant Alexa to represent natural language commands based on an ontology of *actions*, *types*, *properties* and *roles* [22]. Although presenting interesting results, the community has restricted access to their data sets [35].

Although such approach has been proved to be helpful for daily-life situations, it does not attend to the fundamental concept of programming, i.e. to group *sets* of instructions to be repetitively used: function definition.

4   The Typical Task and User

Before discussing the principles and challenges, we need to identify our target users and their typical programming tasks. Blackwell [6] classifies the end users into three groups: those who *like programming*, those who *find programming useful* and those who *believe they will be good at programming*. Among other differences, the groups can be understood from a motivational point of view. While the first and third groups are motivated from an intrinsic desire, moved from an aesthetic vision of programming, the second has a more pragmatic perception, moved by the benefits the technology can offer and evaluating the trade-off of dedicating time and attention to both learning how to program and automating a task. This second group represents the target user profile for NL programming.

Repetitive manual work executed in business context is the typical end-user task we are interested in. Although feasible to be automated, they are neither big enough to call the attention of the IT department nor the company has any IT department at all, such as small ventures and liberal-professional offices. To put in concrete terms, take as an example the following request representing a task attributed to an administrative staff in a hypothetical language school.

> *"Please, send an email to each student from Table 1. Calculate their bills, where each class costs €80, converting the price to their preferred currency. For those in the A1 or A2 levels, please also attach a translated message in their mother languages."*

Processing this task manually for a reasonable long table can demand a few hours of work, although being easily automated if a proper tool and services are available.

5   Principles

Writability, *readability* and *expressivity* have been the main criteria to evaluate programming languages [47]. Contrary to the first thought, we need to resist

Table 1. Tabular content within an email message.

Name	Email	Course level	# of Classes	Currency	Mother language
John	john@smith.com	A1	13	Chinese yuan	Chinese
Kim	kim@korea.com	A2	17	South Korean won	Korean
Tarsila	tarsila@doamaral.br	B2	10	Brazilian Real	Portuguese
...	...	...	...	...	...
Frida	frida@kahlo.com.mx	A2	15	Mexican Peso	Spanish

the temptation of assuming that NL programming would address these issues *by nature*. Indeed, some high-level instructions in natural language look clearer and straightforward, but this is not always the case. For instance, long mathematical expressions described in plain text decrease both readability and writability when compared to the algebraic notation [13]. Data manipulation, an aspect neglected in the mainstream end-user discussions, cannot receive proper support from a pure natural language interface as well [11]. Looking at EUD as more than a language, but as a platform, allows us to overcome these restrictions by using hybrid approaches as we further describe.

Reliability, also commonly accounting for general programming languages evaluation [47], assumes a different sense in the context of NL programming too. As natural language interpreters and parsers need to handle an open lexicon, they always bring a degree of uncertainty [24]. Being reliable in this context means offering to the end user a graceful mechanism to identify and repair any AI misclassifications [40, 46]. For example, when a user types a command and one word is interpreted in a sense different from the user's intention, the output will be incorrect. Any interactive system supported by AI, needs to take into account this risk to provide a *graceful repairing* mechanism.

Moreover, the *information overload* that we currently face cannot be ignored [39]. This phenomenon is not only regarded to books, news, scientific papers and social media posts, but also to *code*. As of 2018, GitHub hosts more than 100 million repositories, each one holding potentially a library [21]. Professional developers already need to make use of specialised blogs and tech social media such as Stack Overflow[1] to find the intended library or API call. Given its nature, any EUD platform must offer suitable mechanisms to help the user to browse and find the pertinent functions.

Despite integrating data be a fundamental step to automate repetitive tasks, EUD platforms frequently miss data manipulation tools [11]. In addition to deal with structured data sources, a new fundamental requirement is the integration of data in the unstructured form, since by 2022, 93% of the information generated is unstructured [20], e.g. texts and also semi-structured data such as Table 1.

---

[1] https://stackoverflow.com.

6   Challenges

An effective EUD tool cannot be only a compiler for a certain syntax but a platform offering three main features. First, the platform needs to allow end users to find and execute both single and composed high-level functions natively integrated with their service providers. Second, it needs to supply tools to deal with unstructured and semi-structured data. And finally, the platform must offer an embedded explanation mechanism.

The failure of the first natural programming languages may be attributed to the attempt to address general-purpose programming, believing they could act as Java or Python [3]. Indeed, as the typical end users think at the level of business concepts and processes, the main advantage of those languages is rather to allow them to operate at higher level of abstraction, connecting coarse-grained pieces of code, a concept well-established in the fields of *Service-oriented Architecture* and *Business Process Management* [37]. For instance, the reference task lists three coarse-grained functions: *send an email, convert currency* and *translate a text*, representing the functions end users see as atomic in that context.

6.1   Information Extraction: The Data

The success of spreadsheets, the most popular EUD platform, can be credited to their ability to fuse data and functions gently [5]. An effective EUD platform needs to have built-in support for data manipulation of complex data types. To a large extent, the unstructured nature of the information we deal with daily is the main barrier to advance the automation of our tasks, since demanding the end users to structure large data sets manually cuts significantly the benefit of automation.

In this context, *information extraction* (IE) becomes a key tool in end-user platforms. Frequently based on artificial intelligence techniques, IE transforms a text into a knowledge graph from which a function can identify and process its content easier [10].

6.2   Search & Run: The Method

In an environment where end-user developers can use services from many sources and providers, finding the appropriate function becomes challenging, a scenario that represents the *information overload* phenomenon for code.

Inspired by feature-rich applications that rely on search to prevent the user from having to navigate in intricate menu hierarchies to find the desired function [8], we advocate a search centred method. Furthermore, given the potential scale of millions of functions, we need to power a search engine with semantic capabilities, especially to deal with the vocabulary gap and ambiguity.

As natural languages allow expressing the same idea using, not only different words but also different syntactic structures, having an efficient search engine is mandatory to disambiguate related services and correctly deal with the vocabulary gap between the way the users express themselves and the way the service is described in the repository. For instance, a user might write *"exchange money"* to find a service that is described as *"convert currency"* [42]. Regarding ambiguity, using natural language command as search queries provides not only an initial description for its functions, but also contextual information that might be used as parameters names or values, and thus serving to disambiguate services sharing similar concepts [40].

After finding the desired function, the end user can run it. The data produced in each execution needs to be shown to the user, but also stored as contextual information to be potentially used by subsequent commands. The end user can then define new functions by chaining a sequence of previously executed commands, consisting of interlinking the output of a command to the input of the subsequent. After receiving a name, the end user can make the new function available to be resolved by the semantic parser. For example, the sequential execution and chaining of *send an email, convert currency* and *translate a text* can generate the new function *"notify student"*, which can be reused with less effort.

Such a method favours a gentle learning curve from two aspects. Novice users find easier to understand stateless programming [23], but to guarantee a pace of learning, the platform needs to allow also a simple transition to write *stateful code*. When using a method based on search, the end user can make this transition naturally, as the same mechanism is used for both the execution of a single command and the construction of functions. Secondly, showing functions in a ranked list dialogues with the principle of *graceful repairing*, reducing the effect of misunderstanding and performance error of the parser [40]. Furthermore, ranked list also allows simple mechanisms to learn from the user experience without any feedback overloading the user.

6.3   Hybrid Interface: The Fusion

None of the interaction modes, i.e. textual language, visual language or programming by example, is able to address alone all expressivity issues of end-user developers. Rather, an effective EUD platform requires multiple mechanisms to tackle appropriately the diverse types of users interactions. Whereas natural languages cover the identification of coarse-grained functions with a semantically-empowered search mechanism, the platform needs to provide proper mechanisms to describe simple computations and to integrate data.

When defining a new function from a chain of commands, as shown in the reference example, the output of one might mismatch (i.e data type or format) to the input of the subsequent. This is the typical duty of scripting languages focused on performing simple computations, such as mathematical expressions and string transformations, as those used in spreadsheets applications. The mathematical formalist in this scenario represents a gain in expressivity [12].

Regarding data integration, both visual tools and programming by example can play important roles. From one side, visual tools can significantly help end users manipulating and understanding the data. Pane and Myers [33] showed that users benefit when data is represented in visual cards that may be traversed manually, or by textual filters. Additionally, it also serves as an important debugging mechanism for the (potentially error-prone) information extraction tools. On the other hand, when importing semi-structured data, comprehending table into texts or content following a standard format, users can benefit from *programming by example*'s techniques with high performance.

6.4   User's Idiosyncrasies: The Human Factor

A tool with the impressive 95% of accuracy fails, in average, every 20 uses, and the impreciseness of intelligent systems can lead to users frustration and lack of interest in the tool [50]. Despite the significant improvement in semantic-based technologies in recent years, the level of complexity inherent to natural language requires that solutions be aware of their (still high) likelihood of failure.

To safeguard user's attention, every AI-empowered functionally needs to provide *graceful repairing* mechanisms to allow proper corrections of misunderstandings in a simple and easy manner. The search approach to identify functions, for example, goes in this direction by giving flexibility to the user to select the correct function among the priorities.

Considering the freedom of expressivity that a natural language allows, the programming model needs to learn the user's writing style. For example, while some users might use complete and grammatically sound sentences to express their commands, others might insist on a keyword-based approach. The same occurs when the user frequently writes a vague word to express a given concept. Learning from the user's history of use is a key component to deliver quality results.

6.5   Explanability: The Transparency

Although much of the debate around explanation for AI systems have been concentrated in decision-making algorithms, mainly guided by the the rights *"for meaningful information about the logic involved"* and *"to non-discrimination"* both defined in the European Union's General Data Protection Regulation [34], other types of AI systems can also benefit from explanations. Biran and Cotton [4] show a set of studies suggesting that users feel much more confident using a system they understand how it works. The recent work of Sales *et al.* [41] reinforced these finds for natural language understanding systems, where users exposed to explanation consistently formed a better mental model, regardless of their technical background. Explanation for AI has become a point of no return for many intelligent systems.

7   Summary

An effective EUD platform is essentially hybrid, combining features from *(i)* natural language programming, *(ii)* visual programming, and *(iii)* programming by example, having, however, a search mechanism with semantic capabilities at its centre. Such a search mechanism focuses on (i) addressing the ambiguity and the vocabulary gap between users and APIs, (ii) supporting a smooth transition from stateless to stateful programming, and (iii) providing graceful repairing capabilities, i.e. an accessible user feedback mechanism to correct or improve automatically generated code. Furthermore, we defend that the platform needs to pay special attention to the manipulation of data. Finally, we highlight the role of explanation in helping users to construct a consistent mental model under the uncertainty associated with artificial intelligence tools.

The new generation of programming platforms for end-user developers has the potential to change how we commercialise, reuse and publish software, as well as creating new business opportunities. Jointly with functionalities such as information extraction and text generation, it can enlarge significantly the spectrum of tasks that can be programmed by end-user developers. The applications emerging from this new context will profoundly affect the level of automation in the society and how humans deal with repetitive tasks.

References

1. Ballard, B.W., Biermann, A.W.: Programming in natural language: "NLC" as a prototype. In: Proceedings of the 1979 Annual Conference, ACM 1979. ACM, New York (1979)
2. Berners-Lee, T., Hendler, J., Lassila, O.: The semantic web. Sci. Am. **284**(5), 34–43 (2001)
3. Biermann, A.W., Ballard, B.W., Sigmon, A.H.: An experimental study of natural language programming. Int. J. Man-Mach. Stud. **18**(1), 71–87 (1983)
4. Biran, O., Cotton, C.: Explanation and justification in machine learning: a survey. In: IJCAI 2017 Workshop on Explainable AI (XAI), p. 8 (2017)
5. Birch, D., Lyford-Smith, D., Guo, Y.: The future of spreadsheets in the big data era. In: Proceedings of the EuSpRIG 2017 Conference "Spreadsheet Risk Management" (2017). ISBN: 978-1-905404-54-4
6. Blackwell, A.F.: End-User Developers – What Are They Like?, pp. 121–135. Springer International Publishing, Cham (2017). https://doi.org/10.1007/978-3-319-60291-2_6
7. Bosca, A., Corno, F., Valetto, G., Maglione, R.: On-the-fly construction of web services compositions from natural language requests. J. Softw. **1**(1), 40–50 (2006)
8. Bota, H., Fourney, A., Dumais, S.T., Religa, T.L., Rounthwaite, R.: Characterizing search behavior in productivity software. In: Proceedings of the 2018 Conference on Human Information Interaction & Retrieval, CHIIR 2018, pp. 160–169. ACM, New York (2018). https://doi.org/10.1145/3176349.3176395
9. Camacho-Collados, J., Pilehvar, M.T.: From word to sense embeddings: a survey on vector representations of meaning. J. Artif. Intell. Res. **63**, 743–788 (2018). https://doi.org/10.1613/jair.1.11259
10. Cetto, M., Niklaus, C., Freitas, A., Handschuh, S.: Graphene: semantically-linked propositions in open information extraction. In: Proceedings of the 27th International Conference on Computational Linguistics, pp. 2300–2311. Association for Computational Linguistics (2018). http://aclweb.org/anthology/C18-1195
11. Desolda, G., Ardito, C., Costabile, M.F., Matera, M.: End-user composition of interactive applications through actionable UI components. J. Vis. Lang. Comput. **42**, 46–59 (2017). https://doi.org/10.1016/j.jvlc.2017.08.004

12. Dijkstra, E.W.: Some comments on the aims of MIRFAC. Commun. ACM **7**(3), 190 (1964)
13. Dijkstra, E.W.: On the foolishness of "natural language programming". In: Program Construction, International Summer School. Springer, London, UK (1979)
14. Empirica: High-Tech Leadership Skills for Europe. Technical report, Empirica (2017)
15. Englmeier, K., Pereira, J., Mothe, J.: Choreography of web services based on natural language storybooks. In: Proceedings of the 8th International Conference on Electronic Commerce, ICEC 2006. ACM, New York (2006)
16. Guida, G., Tasso, C.: NLI: a robust interface for natural language person-machine communication. Int. J. Man-Mach. Stud. **17**(4), 417–433 (1982)
17. Gulwani, S.: Programming by examples (and its applications in data wrangling). In: Verification and Synthesis of Correct and Secure Systems. IOS Press, January 2016. https://www.microsoft.com/en-us/research/publication/programming-examples-applications-data-wrangling/
18. Guzdial, M.: Learner-centered design of computing education: research on computing for everyone. Synth. Lect. Hum.-Cent. Inf. **8**(6), 1–165 (2015)
19. Halevy, A., Norvig, P., Pereira, F.: The unreasonable effectiveness of data. IEEE Intell. Syst. **24**(2), 8–12 (2009). https://doi.org/10.1109/MIS.2009.36
20. IDG: Data & Analytics: Landscape in the Enterprise. Technical report, IDG (2016)
21. Warner, J.: Thank you for 100 million repositories (2018). https://github.blog/2018-11-08-100m-repos/. Accessed 13 May 2019
22. Kollar, T., et al.: The Alexa meaning representation language. In: Proceedings of the 2018 Conference of the North American Chapter of the Association for Computational Linguistics: Human Language Technologies, Volume 3 (Industry Papers), pp. 177–184 (2018)
23. Krishnamurthi, S., Fisler, K.: Programming Paradigms and Beyond. The Cambridge Handbook of Computing Education Research (2019)
24. Li, D., Du, Y.: Artificial Intelligence with Uncertainty. CRC Press, Boca Raton (2017)
25. Lieberman, H., Paternò, F., Klann, M., Wulf, V.: End-User Development: An Emerging Paradigm, pp. 1–8. Springer Netherlands, Dordrecht (2006)
26. Lim, J., Lee, K.H.: Constructing composite web services from natural language requests. Web Semant. Sci. Serv. Agents WWW **8**(1), 1–13 (2010)
27. López, G., Quesada, L., Guerrero, L.A.: Alexa vs. siri vs. cortana vs. google assistant: a comparison of speech-based natural user interfaces. In: International Conference on Applied Human Factors and Ergonomics, pp. 241–250. Springer (2017)
28. Maas, R.E., Suppes, P.: Natural-language interface for an instructable robot. International Journal of Man-Machine Studies **22**(2), 215–240 (1985)
29. Manaris, B.Z., Pritchard, J.W., Dominick, W.D.: Developing a natural language interface for the unix operating system. SIGCHI Bull. **26**(2) (1994)
30. McCann, B., Keskar, N.S., Xiong, C., Socher, R.: The natural language decathlon: multitask learning as question answering. arXiv preprint arXiv:1806.08730 (2018)
31. Miller, G.A.: Wordnet: a lexical database for English. Commun. ACM **38**(11), 39–41 (1995)
32. Namoun, A., Nestler, T., De Angeli, A.: Service composition for non-programmers: prospects, problems, and design recommendations. In: 2010 IEEE 8th European Conference on Web Services (ECOWS), pp. 123–130. IEEE (2010)
33. Pane, J.F., Myers, B.A.: More Natural Programming Languages and Environments, pp. 31–50. Springer Netherlands, Dordrecht (2006)

34. Parliament and Council of European Union: General Data Protection Regulation (EU) no 2016/679 (2016)
35. Perera, V., Chung, T., Kollar, T., Strubell, E.: Multi-task learning for parsing the Alexa meaning representation language. In: Thirty-Second AAAI Conference on Artificial Intelligence (2018)
36. Pop, F.C., Cremene, M., Vaida, M., Riveill, M.: Natural language service composition with request disambiguation. In: Maglio, P.P., Weske, M., Yang, J., Fantinato, M. (eds.) Service-Oriented Computing, pp. 670–677. Springer, Heidelberg (2010)
37. Pourmirza, S., Peters, S., Dijkman, R., Grefen, P.: BPMS-RA: a novel reference architecture for business process management systems. ACM Trans. Internet Technol. **19**(1), 13:1–13:23 (2019). https://doi.org/10.1145/3232677
38. Rodrigues Filho, R., Wild, A., Porter, B.: Code synthesis in self-improving software systems. In: Proceedings of the International Workshop of Self-Improving System Integration. IEEE (2019)
39. Roetzel, P.G.: Information overload in the information age: a review of the literature from business administration, business psychology, and related disciplines with a bibliometric approach and framework development. Bus. Res. (2018). https://doi.org/10.1007/s40685-018-0069-z
40. Sales, J.E., Freitas, A., Handschuh, S.: An open vocabulary semantic parser for end-user programming using natural language. In: 12th IEEE International Conference on Semantic Computing, ICSC 2018, Laguna Hills, CA, USA, January 31–February 2, 2018, pp. 77–84 (2018). https://doi.org/10.1109/ICSC.2018.00020
41. Sales, J.E., Freitas, A., Handschuh, S.: A user-centred analysis of explanations for a multi-component semantic parser. In: Natural Language Processing and Information Systems. Springer International Publishing, Saarbrucken (2020)
42. Sales, J.E., Handschuh, S., Freitas, A.: Semeval-2017 task 11: end-user development using natural language. In: Proceedings of the 11th International Workshop on Semantic Evaluation, Canada (2017)
43. Sales, J.E., Souza, L., Barzegar, S., Davis, B., Freitas, A., Handschuh, S.: Indra: a word embedding and semantic relatedness server. In: Proceedings of the 11th International Conference on Language Resources and Evaluation (LREC 2018). ELRA, Japan (2018)
44. Sammet, J.E.: The use of English as a programming language. Commun. ACM **9**(3), 228–230 (1966)
45. Sangers, J., Frasincar, F., Hogenboom, F., Chepegin, V.: Semantic web service discovery using natural language processing techniques. Expert Syst. Appl. **40**(11), 4660–4671 (2013)
46. Schlegel, V., Lang, B., Handschuh, S., Freitas, A.: Vajra: step-by-step programming with natural language. In: Proceedings of the 24th International Conference on Intelligent User Interfaces, pp. 30–39. ACM (2019)
47. Sebesta, R.W.: Concepts of Programming Languages, 10th edn. Pearson, Upper Saddle River (2012)
48. Sorva, J.: Notional machines and introductory programming education. Trans. Comput. Educ. **13**(2), 8:1–8:31 (2013). https://doi.org/10.1145/2483710.2483713
49. The App Association: Six-Figure Tech Salaries. Technical report, The App Association (2018)
50. Thomason, J., Zhang, S., Mooney, R., Stone, P.: Learning to interpret natural language commands through human-robot dialog. In: Proceedings of the 24th International Conference on Artificial Intelligence, IJCAI 2015, pp. 1923–1929. AAAI Press (2015). http://dl.acm.org/citation.cfm?id=2832415.2832516

51. Vaswani, A., et al.: Attention is all you need. In: Guyon, I., et al. (eds.) Advances in Neural Information Processing Systems 30, pp. 5998–6008. Curran Associates, Inc. (2017). http://papers.nips.cc/paper/7181-attention-is-all-you-need.pdf
52. Wong, J.: Marmite: towards end-user programming for the web. In: IEEE Symposium on Visual Languages and Human-Centric Computing, VL/HCC 2007, pp. 270–271. IEEE (2007)

Improving Accuracy of Automatic Derivation of State Variables and Transitions from a Japanese Requirements Specification

Hiroya Ii[(✉)], Kozo Okano, and Shinpei Ogata

Faculty of Engineering, Shinshu University, 4-17-1 Wakasato, Nagano 380-8553, Japan
16t2007b@shinshu-u.ac.jp, {okano,ogata}@cs.shinshu-u.ac.jp

Abstract. A requirement specification for software is usually described in a natural language and thus may include sentences containing ambiguity and contradiction. Design errors due to ambiguous expressions and contradictions are often found later in the development process. These ambiguous sentences would force the developer to go back to the design process again. In order to prevent this kind of rework, a method of automatically converting a requirements specification written in Japanese to a state transition model is desired to help detect ambiguity and contradiction points of the specification. Our previous research proposed a method to extract the sentence elements necessary for creating a state transition diagram by utilizing morphological analysis techniques and sentence parsing. The extracted elements, however, contained many redundant or unnecessary elements. This paper proposes an improved method to remove such redundant or unnecessary elements by mainly improving parsing rules. This paper also reports the effectiveness of the proposed method by showing the extracted elements necessary for creating a state transition diagram.

Keywords: NLP · Specification · State diagram · Embedded system · CFG

1 Introduction

Most software specifications are written in a natural language, which can result in ambiguous descriptions of what these specifications really mean [1]. Such ambiguity can sometimes make rework of development of software necessary. Ambiguity and incorrectness of specifications are sometimes found by formal approaches including model checking techniques. This method, however, requires specialized knowledge of the model, and is generally difficult to handle. In such a case, formal approaches including model checking tends to be costly as tools for sharing the system specifications between the design team and the development team. Our previous research proposed a method to extract the sentence elements

© Springer Nature Switzerland AG 2020
M. Virvou et al. (Eds.): JCKBSE 2020, LAIS 19, pp. 20–34, 2020.
https://doi.org/10.1007/978-3-030-53949-8_3

necessary for creating a state transition diagram by utilizing morphological analysis techniques and sentence parsing [2]. The method will reduce the cost to use formal and semi-formal approaches. The extracted elements, however, contained many redundant or unnecessary elements. Furthermore, due to the parsing rules are simple, a sentence with a complicated style cannot be recognized as a sentence. This paper proposes a detailed parsing rule set as a method to solve these problems. This paper reports the results of evaluating the effectiveness of a detailed parsing rule. This paper describes our approach and experimental results obtained by applying our method to a specification of an electric pot [3] that is described by SESSAMI.

This paper is organized as follows. Section 2 introduces some past techniques and related work as background. Section 3 and 4 outline our proposed approaches. Sections 5 and 6 present experimental evaluation and discussion, respectively. Finally, Sect. 6 concludes the paper.

2   Background

2.1   Related Techniques

The CYK Algorithm [4] composes grammar trees from a given context-free grammar G and a given sentence s using a dynamic programming algorithm. Usually the grammar G is given in Chomsky-normal form. It requires ($O(n^3)$) calculation time. Figure 1 shows an analysis example of the CYK algorithm. For a sentence "botan wo osu to, mizu wo huttou saseru." Morphological analysis is performed, and each morpheme is identified. For these morphemes and the given typical Japanese grammar, the CYK algorithm is performed. For example, in Fig. 1, the noun "botan" is adjacent to the particle "wo" thus "botan wo" becomes a target phrase. Such analysis is performed for all morphemes. When a sentence is produced the analysis succeeds. Kuromoji [5] is an open source Japanese morphological analysis tool written in Java. Morphological analysis determines the parts of speech of the components of a given text, in other words, the morphemes, which are the minimum units of a text, using its grammar and a dictionary with information on parts of speech. Mecab [6] and JUMAN [7] are famous tools for Japanese morphological analysis. We, however, use kuromoji because of its ease of use with Java.

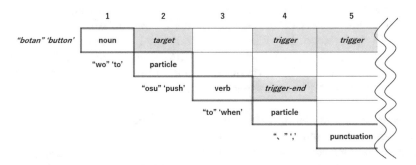

Fig. 1. Parsing a sentence based on a given grammar using the CYK Algorithm.

2.2   Related Work

Natural Language Processing (NLP) is an important technique in information processing. The recent development of NLP has been remarkable [8] and there are many applications of it to various fields. For Japanese texts, KNP [9] and Cabocha [10] are famous tools for syntax analysis. Some of them use support vector machines (SVMs) for processing. Huy and Onishi [11] have proposed a method for checking the correctness of requirements based on ontology and NLP. This group has also studied a method to generate scenarios from a domain specific language (DSL) based on a restricted natural language [12]. Yoshimura and Nakanishi *et al.* [13] have proposed a method based on a Japanese unification grammar to extract simple sentences from technical documents to generate sentences that assist in describing failure modes for HAZOP/FMEA [14]. Our previous works have proposed a method for automatically converting a required specification written in Japanese to a state transition model [2]. This paper is based on the same technique of [2]. We, however, improve the technique in order to improve mainly precision rates for state variables. Compared to other studies, it can be recognized as a sentence even if it is described in various expressions.

3   Proposed Method

Our proposed method consists of the following six steps. Each step is performed on a sentence, therefore our proposed method does not deal with context of sentences.

1. Perform morphological analysis on required specifications written in Japanese using existing tools;
2. Perform parsing using the CYK method;
3. Output the syntax analysis result in XML;
4. Classify outputs;
5. Extract variables and variable-values; and
6. Extract object and action.

3.1   Morphological Analysis

In the case of spoken Japanese, listeners can understand ambiguous and incomplete sentences. In the case of written Japanese, however, such sentences it may be wrongly understood. It is also difficult to be analyzed automatically. For the revised sentences, we analyze each of the sentences using kuromoji. The result is output to an XML file. Using the user dictionary provided by kuromoji library, we can analyze several domain-specific words with specified parts of speech and morphemes.

Therefore, we manually extracted sentences from the requirements specification, and rewrite sentences containing unnatural expressions. For example, a sentence, "Mansui sensa ha, deforuto ha, on de aru." (Translated literally,"For a full water detection sensor, the default is on.") should be rewritten. The problem with the sentence is that the subject, "Mansui sensa," ("full water detection sensor") which modifies "deforuto," (translated literally, "default") uses the same expression as the nominative. For example, in this statement, we change to "Mansui sensa no, deforuto ha, on de aru." ("For a full water detection sensor, the default is on.")

3.2   Parsing

The major difference between the previous study and this paper is whether the rules of parsing are detailed according to the characteristics or the mode of the description of the required specifications. In previous research, sentences are analyzed by simple syntax rules, and the elements necessary to create a state transition diagram were extracted. There are two problems with the parsing rules of the previous research.

1. One phrase "noun phrase" has multiple meanings.
2. Small stylistic changes make it unrecognizable as a sentence.

Problem 1 appears in the following sentence. For example, the sentence "futto botan wo osuto, potto nai no mizu wo futto sa se te karuki nuki wo okonai masu." ("When the boiling button is pressed, the water in the pot is boiled and desalted.") is analyzed using the parsing rules of the previous research, the two phrases "futto botan wo" ("Boiling button") and "potto nai no" ("in the pot") are recognized as the same "noun phrase." The "futto botan" is a phrase that indicates the target of the operation, and "potto nai no" is a possessive that modifies the following "water." In other words, each phrase has a different role in the sentence. In this example, the phrase, "futto sa se te" ("be boiled, and") is the entry operation in the state transition diagram. The object to be boiled is "potto nai no mizu." Therefore, the phrase "potto nai no mizu wo," ("to water in the pot") includes elements that need to be extracted to create a state transition diagram. The previous study, however, recognizes "potto nai no mizu wo" as a noun phrase without recognize "potto nai no mizu" as a noun. When this parsing rule is used, we analysis that "potto nai no" (noun phrase) followed by "mizu wo" (noun phrase) as "potto nai no mizu wo" (noun phrase). As a result, the previous method cannot extract the noun phrase "potto nai no" and the noun "mizu" ("water") from the noun phrase "potto nai no mizu wo" which is composed of the noun phrase and the noun phrase. In addition, in this situation, there is a problem that in order to force extraction, it is necessary to use extraction rules which are not generic in the next step.

Problem 2 occurs when the ending of the phrase indicating the condition changes. In the case of descriptions about conditions written in Japanese, the following three descriptions all have the same meaning. "kyuuto chu de aru baai," "kyuuto chu de aru baai," "kyuuto chu de aru baai ni." The second and third descriptions are one-letter changes at the end of the first description. In the second sentence, we removed " ',','" and in the third sentence, we change "," to "ni." Using the parsing rules of previous research, the first description could be recognized as a phrase related to the condition, but the other description could not be recognized as a phrase related to the condition. In the parsing rules of previous research, this problem occurred because phrases related to conditions were analyzed using "," and a specific word ("baai," "toki" etc.) representing the condition as a mark. In the following, we describe the new parsing rules and solution ideas proposed in this paper.

Problem 1 is that one phrase type has multiple meanings. In this paper, we created new phrase types. In addition, we created parsing rules for the phrase types. For example, if the noun followed by a particle "no," it is considered a possessive. When the noun followed by a particle "ha," it is considered a nominative. If a noun follows a possessive, it is considered a noun. When parsing the phrase "potto nai no mizu" using the parsing rules created in this paper, we analysis that "potto nai no" (possession) followed by "mizu" (noun) as "potto nai no mizu" (noun). We analysis that "potto nai no mizu" (noun) followed by "wo" (particle) as "potto nai no mizu wo" (operation target). When extracting "potto nai no mizu" from the sentence analyzed, we can extract this phrase using the condition "all of the elements in the sub tree without particles in the operation target child node." Using this condition, it is possible to accurately extract nouns that indicate the target of action, not only specific sentences but general sentences.

Problem 2 is that there are few parsing rules for phrases that function on conditions. In this paper we created new parsing rules for phrase related to condition. We show part of the parsing rules we create in Table 1. The phrase "kyuto chu" is analyzed as a noun. The phrase "de aru" is analyzed as "definition-end," because this phrase is applied rule 1 in Table 1. "de aru baai" is analyzed as "definition-trigger-end," because this phrase is applied rule 2 in Table 1. Using new parsing rules, we analyze the second description "kyuuto chu de aru baai" as a phrases related to conditions, because this phrase is applied rule 5 in Table 1. We also analyze the first description's phrase "de aru baai," as "definition-trigger-end," because this phrase is applied rule 3 in Table 1. After that, rule 5 in Table 1 is applied, and the phrase "kyuuto chu de aru baai," is also analyzed as a phrases related to conditions. We also analyze the third description's phrase "de aru baai ni" as "definition-trigger-end," because this phrase is applied rule 4 in Table 1. After that, rule 5 in Table 1 is applied, and the phrase "kyuuto chu de aru baai ni" is also analyzed as phrases related to conditions. Therefore, when using this parsing rule, we can analyze three phrases "kyuuto chu de aru baai," "kyuuto chu de aru baai" and "kyuuto chu de aru baai ni" as phrases related to conditions.

We analyze an example sentence "botan wo osu to, mizu wo huttou saseru." to make a comparison between conventional method and proposed method. The sentence is used only for explanation, and it is not a target for the experiment. We show the result that is analyzed using the conventional method in Fig. 2.

In addition, we show the result that is analyzed using the proposed method in Fig. 3. We can see Problem 1 in Fig. 2. The phrase "botan wo osu to" contains meaning of sentence continuation, however "botan wo osu" does not. The phrases "botan wo osu" and "botan wo osu to" are, however, analyzed as same "simple sentence" using the conventional method. By analyzing this way, it is difficult for extracting elements using XPath. In the other hand, we can analyze "botan wo osu" and "botan wo osu to" as different kinds of phrases using proposed method. As shown in Fig. 3, "botan wo osu" is not analyzed. The phrases "botan wo" and "osu to" are, however, analyzed as "target" and "trigger end", respectively. Finally, we can analyze "botan wo osu to" as "trigger". By analyzing this way, we can easily extract element like target of motion in trigger using XPath. Also, Problem 2 exists as shown in Fig. 2. The phrase "botan wo osu to" is analyzed as "simple sentence", however "botan wo osu to," is analyzed as "simple sentence-contd." The word "contd" in "simple sentence-contd" is shorthand of "continued." We analyze "simple sentence-contd" and following "simple sentence" as "sentence." In brief, if we use conventional method, we must rely on punctuation to analyze continuous expressions in sentences. We cannot recognize "sentence" without punctuation in this sentence. In addition, both of "botan wo osu to" and "botan wo osu to," appear as continuous expression in Japanese writing sentence. Therefore, these expressions should be analyzed as the same phrase and we should not rely on punctuation. We can analyze "botan wo osuto" and "botan wo osuto," as same "trigger" using proposed method as shown in Fig. 3.

Table 1. Example of new parsing rules

	Seq1	Seq2	Result
1	"de"	"aru"	definition-end
2	definition-end	"toki", "baai"	definition-trigger-end
3	definition-trigger-end	","	definition-trigger-end
4	definition-trigger-end	"ni"	definition-trigger-end
5	noun	definition-trigger-end	a-trigger-end

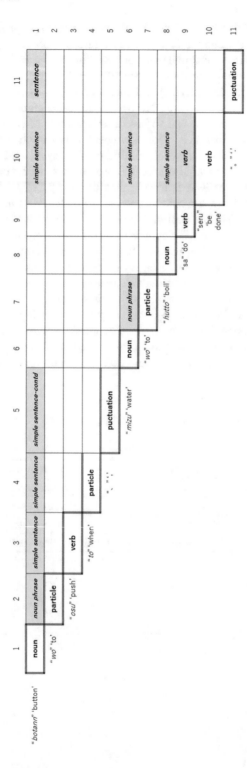

Fig. 2. Conventional method

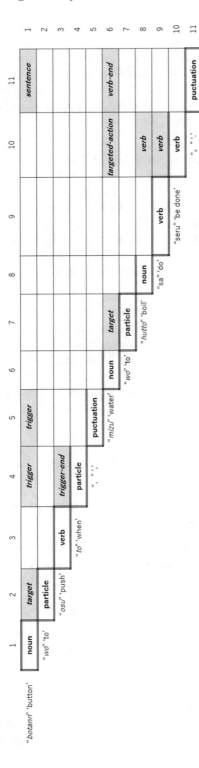

Fig. 3. Proposed method

3.3   Classification

In this study, a sentence which defines something, is called a "definition" sentence, and a sentence which describes the processing to be performed when a certain condition is satisfied, is called a "processing" sentence. We classify each sentence into definition sentences, processing sentences and others. Using the information on the classification, we choose the elements to extract [15].

Conditions that Supports a "definition" Sentence. If a sentence contains a nominative phrase or a supplement phrase, it is recognized as a definition sentence. The noun phrase is noun followed by "ha." The supplement phrase is noun followed by "de," or noun followed by "ni." If these expressions are followed by " ',' " our method recognizes the phrase as a "condition."

(ex.) "Suii sensa ha, suii wo kenshutu suru." ("The water level sensor detects the water level.")

"Suii sensa ha" is a nominative phrase. The sentence is a definition sentence.

Conditions that Supports a "processing" Sentence. If the sentence has terminal expressions of conditional phrases, it is recognized as a processed sentence. The basic terminal expressions of conditional phrases are "toki," "to" and "baai,." If this expression is followed by "," or "ha," our method recognizes as phrase related to conditions.

(ex.) "Rokku kaijo joutai de aru toki, kyuutou dekinai." ("Hot water cannot be supplied when unlocked.") "Rokku kaijo joutai de aru toki," is a condition. The sentence is a processing sentence.

3.4   Extraction of Operation of Condition and Target of the Operation

Using our defined rules based on XPath, the elements are extracted as follows.

Table 2. Results of Experiment 1

	Subject	Operation1	Target1	Operation2	Target2
Number of corrects	20	11	10	11	10
Number of extractions	20	13	12	13	12
Recall	1.00	1.00	1.00	1.00	1.00
Precision	1.00	0.84	0.84	0.84	0.83

Subject in definition sentence ("Subject" in Table 2) We extracted the object to be defined. For example, "Suii sensa ha suii wo, kenshutsu suru," we extract "Suii sensa" ("water level sensor") which is a nominative of the sentence.

Operation of condition in processing statement ("Operation1" in Table 2) For example, for a sentence "Pot ga kyuuto chu de aru toki, rokku dekinai," we extract "kyuuto chu de aru" as an operation.

Operation target of condition and operation in processing statement ("Target1" in Table 2) For example, "Pot ga kyuuto chu de aru toki, rokku dekinai," we extract "pot" which is a target of the operation "kyuuto chu de aru."

Rules for extracting the operation of the condition in the processing statement are described in XML as shown in Listing 1.1. The common tag specifies target to be searched. In this case, the description of "// sentence [@ decision = 'processing' or @ decision = 'processing, definition'] // phrase [@ type = 'a-trigger-end']" is search starting point (the root node of the tree). Similarly, a phrase "a-trigger-end" is a phrase that combines the ending of the definition sentence "de aru" and the definition target, followed by the ending of the conditional phrase. Therefore, the search target is "pot ga kyuutou chu de aru toki" ("when pot supplies hot water.") The match tag describes the condition to be extracted. In this example, the third and fourth conditions from the top of the figure are met. "kyuuto chu" ("when hot water is being supplied") is analyzed as a phrase meaning a noun in which "kyuuto" ("hot water is being supplied") and "chu" ("when") are combined. Therefore, the fourth condition, "the phrase element meaning all nouns in the search target" is satisfied and "kyuuto chu" is extracted. Since "de aru" is analyzed as the ending of the definition sentence, it is extracted by satisfying the third condition "all the ending of the definition sentence in the search target."

Listing 1.1. Example of XPath

```
<rule>
  <common expression="//sentence[@decision='processing' or @decision
      ='processing, definition']//phrase[@type='a-trigger-end']"/>
    <match decision="variable-value" expression=".//phrase[@type='
        trigger-end']/phrase[@type='verb']//token"/>
    <match decision="variable-value" expression="./token"/>
    <match decision="variable-value" expression=".//phrase[@type='
        definition-end']/token"/>
    <match decision="variable-value" expression=".//phrase[@type='
        noun']//token"/>
</rule>
```

3.5 Extraction of Operation and Operation Target When Conditions Are Satisfied

Using our defined rules based on XPath, the elements are extracted as follows.

Operation target of the operation when the condition in the processing statement is satisfied ("Operation2" in Table 2) For example, for a sentence, "Pot ga kyuuto chu de aru toki, rokku dekinai." "rokku," which is the object of "dekinai," is extracted.

Operation when the condition in the processing statement is satisfied (**"Target2" in Table** 2) For example, for a sentence, "Pot ga kyuuto chu de aru toki, rokku dekinai." "dekinai," which is the operation of "rokku," is extracted.

4   Improved Method

In this paper, we further refine the parsing rules in order to create a state transition diagram. The steps in Sects. 4 are the same as the steps 3.1 to 3.3. However, the contents of the extraction differ in the subsequent steps. This section describes only the processes that are different from those in Sect. 3.

The extraction method uses rules using XPath described in XML in the same way as the "method" described in Sect. 3.

4.1   Extraction of State and Condition Before Transition

We extract the element using XPath. The elements to be extracted are as follows.

State Before Transition. We extract state nouns. Here, a state noun is a noun followed by "tyuu" or "joutai." However, state nouns in the transition phrase are excluded because they represent the state after the transition. For example, "hutto chu ni hoon botan wo osuto, hutto wo tyuusi si te, hoon joutai ni naru" ("When the boiling button is pressed during boiling, the boiling is stopped and the state shifts to keeping warm state."), "hutto chu" ("during boiling") is extracted as the state before the transition.

Condition. We extract transition conditions. In the conditional phrase, we extract the contents other than the conditional phrase ending with "to." For example, "Ondo seigyo kanou na suii de aru toki, hutto joutai ni iko suru." ("When the water level is at a temperature controllable, the state shifts to a boiling state."), "Ondo seigyo kanou na suii de aru" ("The water level is at a temperature controllable") is extracted as a condition.

4.2   Extraction of State After Operation and Transition

Operation. We extract the operation that trigger of the transition. In the conditional phrase, we extract the contents of the conditional phrase ending with "to." For example, "hutto chu ni hoon botan wo osuto, hutto wo chuusi si te, hoon joutai ni naru," "hoon botan wo osuto" ("pressing the boiling button") is extracted as operation.

State After Transition. We extract state nouns. However, only state nouns in the transition phrase is extracted. For example, "hutto chu ni hoon botan wo osuto, hutto wo chuusi si te, hoon joutai ni naru.," "hoon joutai" ("warm state") is extracted as the state after the transition.

5   Evaluation

This paper has the following two research questions.

RQ1: "Is the detailed parsing rule proposed in Sect. 3 valid?"
RQ2: "Whether the proposed method improves the results of the conventional method?"

5.1   Evaluation Experiment and Evaluation Criteria for RQ1

The evaluation experiment for RQ1 (Experiment 1) uses the description of the requirement specification about an electronic pot (7 Edition) published by SESSAME. We manually determined the elements that meet the conditions in Sects. 3.4 and 3.5 as correct answer. Experiment 1 verifies whether parsing the syntax in consideration of the meaning of clauses may lead to generalization of the extraction rules. The operations of the conditional phrases and the elements of the correct set of the operation targets are only the first conditional phrases in the sentence to be analyzed. Similarly, if there are multiple operations when the conditions are satisfied, the operation described last is determined as the correct answer. We set a total of 31 sentences as the sentences including the extraction target. We use recall and precision as evaluation criteria.

5.2   Evaluation Experiment and Evaluation Criteria for RQ2

In the evaluation experiment (Experiment 2) for RQ2, a state transition diagram was created using astah [16] while consulting with four UML beginners based on the requirement specification about an electronic pot (6 Edition) published by SESSAME. It is derived mainly using the descriptions in Sect. 7 of the requirements specification, but also referred to the descriptions in Sects. 3, 5, and 6 of the requirements specification. In this way, a state transition diagram was obtained as the correct answer for this experiment. A total of 36 natural language sentences were targeted. We use recall and precision as evaluation criteria.

5.3   Results of Experiment 1

Table 2 shows the experimental results. The recalls are 1.00 for the four extracted elements. We can extract all the elements designated as correct answer. The precision is a high score of higher than 0.80 in the four types of extracted elements. However, extra elements have been extracted.

5.4   Results of Experiment 2

Table 3 shows the experimental results. The recall rates are low for all four extracted elements. The precision is 1.00 for all four extracted elements. We extract a lot of elements designated as correct answer. Although it could not be output, it is able to extract accurately.

Table 3. Results of Experiment 2

	State before transition	Condition	Operation	State after transition
Number of corrects	3	11	13	6
Number of extractions	1	2	1	3
Recall	0.33	0.18	0.07	0.50
Precision	1.00	1.00	1.00	1.00

5.5 Summary

From the experimental results, we can answer to two research questions like this as follows. RQ1: Refinement of parsing rules is effective for improving the accuracy of element extraction. RQ2: The proposed method improves the results of the conventional method.

6 Discussion

6.1 Discussion of Experiment 1

The purpose of the experiment 1 is to examine how detailed analysis of the parsing rules of the requirements specification written in natural language would affect the analysis results. As a result, recalls are all 1.00. We can extract the target element is without leaking. The reason why the precision is not 1.00 is obvious. The parsing rules used in this study set rules that can be analyzed as sentences even if there are multiple conditions and operations corresponding to conditions. If we use an XPath predicate and specify only the first phrase, the precision will all be 1.00. Therefore, it is verified that the recall and precision of the research following the previous research could be improved by refining the parsing rules.

6.2 Discussion of Experiment 2

Experiment 2 shows the usefulness of the improvement in Sect. 4. In the improvement, we added new rules to the rules of parsing proposed in Sect. 3 for extracting necessary elements in the state transition diagram. Making the parsing rules more detailed and giving meaning to each phrase, lets the extraction rules contribute to improve the extraction accuracy. Most of the sentences analyzed in Experiment 1 had simple structures which is composed of conditions and the processing when the conditions are satisfied. We, however, could not analysis many of the sentence used in Experiment 2 could not be analyzed because of their complex structure. As a result of the parsing used in this paper, we can give only the name of the phrase. We consider that it is possible to set more flexible parsing rules specifying additional information in the phrase in the results of parsing. "state before transition" and "state after transition" may often be omitted in the description of the actual requirement specification. When creating a state transition diagram by hand operation from a requirement specification composed of such sentences. it is often the case that the state transition is determined by considering the context of the description. Therefore, it also may be possible to cope with the lack of description about states in automatic generating, by making changes to the program that extracts elements from sentences so that the state transitions of the entire requirement specification can be retained.

7   Conclusion

In this paper, we proposed a refinement of parsing rules as a method of extracting elements necessary for creating a state transition diagram from specification described in natural language. Furthermore, we applied the proposed method to the required specifications and performed experiments. In the experiments, we examined how the parsing rules newly proposed is improved than the parsing rules of the previous research. As a result, improvement of extraction accuracy was confirmed by refinement of the parsing rules. In the method of Sect. 4, we modified the syntax rules created in the method of Sect. 3, for extracting the elements required to create a state transition diagram. However, because the parsing rules were not sophisticated, they often could not be recognized some inputs as sentences. On the other hand, in some cases, the elements necessary to create a state transition diagram can be extracted from input sentences which can be recognized as sentences by the proposed method.

As future work, first, it is required to refine the parsing rules specialized for the requirement specification written in natural language. In order to set more flexible parsing rules, it is necessary to consider using parsing methods other than the CYK method.

Acknowledgement. Part of this research is supported by Grants-in-Aids C16K00094 and 18H04094.

References

1. Bertrand, M.: On formalism in specifications. IEEE Softw. **2**(1), 6–26 (1985)
2. Okano, K., Takahashi, K., Ogata, S., Sekizawa, T.: Analysis of specification in Japanese using natural language processing. In: Proceedings of the 12th Joint Conference on Knowledge-Based Software Engineering (JCKBSE 2018), pp. 12–21. Springer, Cham (2018)
3. SESSAMI: Embedded system educational material 'Wadai-Futto pot GOMA1015 type' required specification. http://www.sessame.jp/workinggroup/WorkingGroup2/POTSpecication.htm (in Japanese)
4. Kasami, T.: An efficient recognition and syntax-analysis algorithm for context-free languages. Scientific report AFCRL-65-758. Air Force Cambridge Research Lab, Bedford, MA (1965)
5. Atilica, kuromoji. https://www.atilika.com/ja/kuromoji/. Accessed 1 Mar 2020

6. MeCab: yet another part-of-speech and morphological analyzer. http://taku910. github.io/mecab/. Accessed 24 Feb 2020
7. JUMAN. http://nlp.ist.i.kyoto-u.ac.jp/index.php?JUMAN. Accessed 24 Feb 2020
8. Nihar, Ranjan, Kaushal, Mundada, Ahmad, Saim, Phaltane, Kunal: A survey on techniques in NLP. Int. J. Comput. Appl. **134**(8), 6–9 (2016)
9. Sasano, R., Kurohashi, S.: A discriminative approach to Japanese zero anaphora resolution with large-scale lexicalized case frames. In: Proceedings of the 5th International Joint Conference on Natural Language Processing (IJCNLP 2011), pp. 758–766 (2011)
10. Kudoh, T., Matsumoto, Y.: Japanese dependency analysis based on support vector machines. In: Joint SIGDAT Conference on Empirical Methods in Natural Language Processing and Very Large Corpora EMNLP/VLC 2000 (2000)
11. Huy, B.Q., Ohnishi, A.: A verification method of the correctness of requirements ontology. In: Proceedings of the 10th Joint Conference on Knowledge-Based Software Engineering (JCKBSE 2012), pp. 1–10 (2012)
12. Ohnishi, A., Kitamoto, K.: A generation method of alternative scenarios with a normal scenario. IEICE Trans. Inf. Syst. **E93-D**(4), 693–701 (2010)
13. Yoshimura, K., Nakanishi, T., Ototake, H., Tanabe, T., Furusho, H.: Extraction of simple sentences from technical documents using Japanese unification grammar. IEICE Technical report, KBSE2017-58, vol. 117(465), pp. 109–114 (2018) (in Japanese)
14. IEC 61882:2001: Hazard and operability studies (HAZOP studies) - Application guide. Accessed 1 Mar 2020
15. Yumikura, Y., Wada, T., Sumi, T., Fujinmoto, H., Murata, Y.: Applying natural language analysis to the evaluation of requirements specifications. IPSJ Technical report, 2013-SE-181, pp. 1–8 (2013) (in Japanese)
16. Change Vision: Astah-professional. http://astah.net/products/. Accessed 1 Mar 2020

Supporting Change Management
of Sequence Diagrams

Toshitaka Odamura, Takayuki Omori, and Atsushi Ohnishi[✉]

Ritsumeikan University, Kusatsu 5158577, Shiga, Japan
{omori,ohnishi}@cs.ritsumei.ac.jp

Abstract. UML Sequence diagram is an effective model to specify behaviors of the target system. Sequence diagrams can be used for specifying interactions between a system and its user, and interactions between objects. Just like scenarios, sequence diagrams are useful for requirements elicitation in the requirements definition. Sometimes sequence diagrams should be modified in accordance with requirements changes or correction of errors. In such a modification, careless mistakes may lead to include another error in the diagram. As a result, careless modifications may lead to a failure of software development. In this paper, we propose a method of visualization of the difference between original sequence diagram and modified one by emphasizing the difference in order to eliminate newly produced errors in the modified sequence diagram. We adopt PlantUML as a description language and a viewer of sequence diagrams. We illustrate our method using a prototype system written with Java with an example.

Keywords: Sequence diagram · Change management of sequence diagrams

1 Introduction

Sequence diagrams are widely used for specifying interactions among systems, subsystems, objects and users. In a software development, sequence diagrams should be modified when requirements are changed or some errors in the diagrams are found. After modification of the diagram, developers who did not modify the diagram may not grasp the difference between the original diagram and modified one.

If the diagram was wrongly modified and includes some new errors due to the wrong modification, the software development will be damaged and may be failed.

In order to avoid this problem, we propose a method of (1) visualizing the difference between the original diagram and the modified diagram and (2) producing a report that summarizes the difference in order that developers who did not modify the diagram can easily understand the difference.

© Springer Nature Switzerland AG 2020
M. Virvou et al. (Eds.): JCKBSE 2020, LAIS 19, pp. 35–46, 2020.
https://doi.org/10.1007/978-3-030-53949-8_4

In Sect. 2, we briefly introduce sequence diagram and PlantUML [7]. In Sect. 3, related works will be described. In Sect. 4, we will show our method of visualization of the difference and production of summarized report. In Sect. 5, a prototype system based on the method will be illustrated with an example. In Sect. 6, we will conclude our research.

2   Sequence Diagram

Sequence diagrams specify sequential interactions among objects and actors. A sequence diagram consists of title, objects, messages, combined fragments and execution specifications. Figure 1 shows a sequence diagram of cash account system.

In Fig. 1, the title of the diagram is shown in the first line as "cash account system." There exist an actor named "customer" and two objects named "clerk" and "cashier." Customer who wants to purchase items brings them to clerk. Clerk checks the items one by one. In case of liquor, the clerk checks the age of customer. Without problems, the clerk scans the bar code of the item. There exist three combined fragments. The most outside fragment is "loop." The clerk checks each item and scans its bar code until all of the items are processed. The second fragment is "opt." This fragment means conditional branch. In this case, if item is liquor, the clerk check the age of the customer. The most inside fragment is "alt." This fragment means if-then-else. In this case, if the customer's age is less than 20, the clerk refuses to sell the liquor, but in the else case he can sell it. After exit of the loop, the cashier object sends "total amount" to the customer via the clerk. The customer pays the amount to the cashier via clerk. The first arrow in the diagram means a message named "items" from the customer to the clerk. Vertical boxes mean execution occurrences corresponding to execution specifications.

We adopt PlantUML as a description language of sequence diagrams. Figure 2 shows the description of the sequence diagram shown in Fig. 1 with PlantUML. The line numbers are not necessary, but we give them for users' convenience.

In Fig. 2, the activate statement means that the corresponding object becomes instantiated, while the deactivate statement means that the object becomes released. The destroy statement means that the object stops its life.

3   Related Works

Triandini et al. proposed a measurement similarity method between sequence diagrams [10]. They used the cosine similarity of two vectors each of which corresponds to a sequence diagram. The elements of vector are objects and messages. They can calculate the similarity between two sequence diagrams, but they cannot show the different elements of the two sequence diagrams.

There exist several approaches to verify the consistency between sequence diagram and another model, such as state machine diagram, usecase diagram, and class diagram [2, 6, 9, 11].

Matsumoto et al. proposed a method of checking the consistency between sequence diagrams and state machine diagrams using FDR model checker [4]. Their approaches cannot be applied to analyze the difference between sequence diagrams.

In [8], a method of comparing between two sequence diagrams is presented. This method enables to detect the longest common message passings (they say "duplication") [3], but cannot detect added, deleted , or updated message

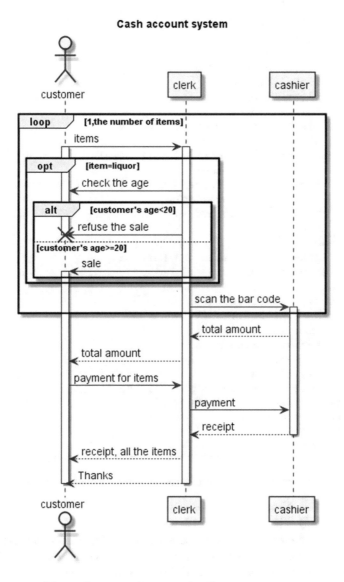

Fig. 1. Sequence diagram of cash account system.

passings. In contrast, our method can detect not only duplication, but also added, deleted, or updated message passings. Moreover, our method can detect different elements, such as objects, message labels, combined fragments and so on, between two sequence diagrams.

```
1)  @startUML{cashier.png}
2)  title Cash account system
3)  actor customer
4)  participant clerk
5)  participant cashier
6)  loop 1,the number of items
7)    customer->clerk:items
8)    activate customer
9)    activate clerk
10)   opt item=liquor
11)     clerk->customer:check the age
12)     alt customer's age<20
13)       clerk->customer:refuse the sale
14)       destroy customer
15)     else customer's age>=20
16)       clerk->customer:sale
17)       activate customer
18)     end
19)   end
20)   clerk->cashier:scan the bar code
21)   activate cashier
22) end
23) cashier-->clerk:total amount
24) clerk-->customer:total amount
25) customer->clerk:payment for items
26) clerk->cashier:payment
27) cashier-->clerk:receipt
28) deactivate cashier
29) customer<--clerk:receipt, all the items
30) customer<--clerk:Thanks
31) deactivate clerk
32) deactivate customer
33) @endUML
```

Fig. 2. Sequence diagram of cash account system written with PlantUML.

Mergely checks the difference between two text files [5]. Though there exist surficial differences that are not semantical differences, Mergely detects differences in these two lines. For example, "customer->clerk:message" and "customer->clerk : message" in PlantUML files are semantically the same, but Mergely detects that they are different, because "->" and ":" are not delimiters for Mergely.

Astah professional shows a report of the difference between two sequence diagrams. It detects different types of objects and added/deleted objects [1]. Since Astah professional cannot directly show the difference of sequence diagrams, it may be difficult for developers to understand the difference between two sequence diagrams. Moreover, it cannot detect the difference between contents of combined fragments and the difference between messages.

Our aim is to detect the difference between combined fragments, messages and not to detect surficial differences. In other words, we do not detect the difference if they are syntactically different, but semantically the same each other. For example, if the order of object is changed in a sequence diagram, the layout of the changed diagram becomes different with that of the original diagram. However, the changed diagram is semantically the same as the original one. So, we do not detect the difference between them in such a case.

We also show the two sequence diagrams in which different elements of the diagrams are colored in order to easily understand the differences between them.

4   Differential Analysis Method of Sequence Diagrams

For given two PlantUML files of sequence diagrams, we assume one file as a previous diagram, and the other as a new diagram. We detect the same elements, updated elements, added elements, and deleted elements between two diagrams, where elements mean objects, message passings, combined fragments, or execution specifications. Our detection method depends on the elements, that is, objects, message passings, and combined fragments as shown in below.

4.1   Differential Analysis of Objects

In definition of an object, (1) object type, (2) object name, and if exist, (3) its alias(es) are the same between two files, these objects are the same. Object types are actor, participant, control, entity, collections, and database. As for objects that have no the same objects, if there are object definitions where one of the three are different, but the others are the same, we assume that these are corresponding objects each other, and this object in the new diagram is updated. Including these cases, Table 1 shows the similarity of two objects between two diagrams.

In Table 1 "-" shows same or different or none (not specified.) Not same or updated objects in the new diagram are added objects and such objects in the previous diagram are deleted objects in the new diagram.

In Fig. 3, "actor Foo1," "entity Foo4," "database Foo5," "collection Foo6" are the same objects. They correspond to the case no. 1 in Table 1. "Boundary foo2" in the left is updated, because object name in the left is the same as alias in the right as shown in case no .2. "Control Foo3" in the left is updated, because object names are the same, but aliases are different (case no. 3). "Participant Foo7" is updated, because object types are different (case no. 5). "Participant Foo8" is deleted and "participant Foo9" is added.

Table 1. Difference between objects

Case no.	Type	Name	Alias	Object
1	Same	Same	Same or not specified	Same
2	Same	Different	Same	Updated
3	Same	Same	Different	Updated
4	Same	Different but same as alias	-	Updated
5	Different	Same	Same or not specified	Updated
6	Different	Same	Different	Different
7	Same	Different	Different or not specified	Different
8	Different	Different	-	Different

```
actor Foo1
boundary foo2
control Foo3 as foo3
entity Foo4
database Foo5
collections Foo6
participant Foo7
participant Foo8
...
```

```
actor Foo1
boundary Foo2 as foo2
control Foo3 as foo
entity Foo4
database Foo5
collections Foo6
actor Foo7
participant Foo9
...
```

Fig. 3. Difference between object definitions.

4.2 Differential Analysis of Message Passing

One message passing consists of (1) a sender object, (2) a receiver object, and (3) a message between them. If there exists just one message passing whose three elements are the same between two diagrams, we assume that these two message passings are the same. If there exist multiple message passings whose three elements are the same between two diagrams, our tool asks a software designer which message passings are corresponding to between two diagrams. If there exist message passings whose two elements are the same, but one element is different or updated each other, our tool also asks the designer which message passings are corresponding to between two diagrams. After the designer's decision, these two message passings will be corresponding to or not. Otherwise, a message passing in the previous diagram does not have any corresponding message passing in the new diagram becomes deleted in the new diagram. A message passing in the new diagram does not have any corresponding message passing in the previous diagram becomes added in the new diagram.

In Fig. 4, definitions of object are the same as the definitions in Fig. 3. "Foo1 -> foo2 : To boundary" is the same, because the senders and message labels are the same, and receivers are different, but alias names are the same between the two diagrams. "Foo1 -> Foo3 : To control" is updated, because the senders

```
actor Foo1
boundary foo2
control Foo3 as foo3
entity Foo4
database Foo5
collections Foo6
participant Foo7
participant Foo8
Foo1 -> foo2 : To boundary
Foo1 -> Foo3 : To control
Foo1 -> Foo4 : To entity
foo2 -> Foo8 : To participant
...
```

```
actor Foo1
boundary Foo2 as foo2
control Foo3 as foo
entity Foo4
database Foo5
collections Foo6
actor Foo7
participant Foo9
Foo1 -> Foo2 : To boundary
Foo1 -> Foo3  : To control
Foo1 -> Foo4 : To entity
Foo2 -> Foo9 : To participant
...
```

Fig. 4. Difference between message passings.

and message labels are the same, but alias names of the receivers are different between the two diagrams. Message passing, "Foo1 -> Foo4 : To entity" is the same, because sender, message, and receiver are the same between two diagrams. "foo2 -> Foo8 : To participant" is deleted in the new diagram, because two elements of the message are not the same. "Foo2 -> Foo9 : To participant" is added in the new diagram, because two elements of the message are not the same.

4.3 Differential Analysis of Combined Fragment

A combined fragment consists of (1) kind of operator, (2) guard, and (3) statements, where statements include message passings, execution specifications, and combined fragments. Operator defines type of combined fragment. Table 2 shows operators.

Table 2. Kinds of operator

Operator
alt, opt, par, loop, critical, neg, assert, strict, seq, ignore, consider

If there exists just one combined fragment whose three elements are the same between two diagrams, we assume that these two combined fragments are the same. If there exist multiple combined fragments whose three elements are the same between two diagrams, our tool asks a software designer which combined fragments are corresponding to between two diagrams or not.

If there exists a combined fragment whose operator and some statements are the same between two diagrams, we ask a software designer whether these two combined fragments are corresponding to or not.

```
. . .
loop 1,the number of items
   customer->clerk:items
   activate customer
   activate clerk
   opt item=liquor
      clerk->customer:check age
      alt customer's age<20
         clerk->customer:refuse
         destroy customer
      else customer's age>=20
         clerk->customer:sale
         activate customer
      end
   end
   clerk->cashier:scan bar code
   activate cashier
end
. . .
```

```
. . .
loop 1,the number of items
   customer->clerk:items
   activate customer
   activate clerk
   opt item=tobacco
      clerk->customer:check age
      alt customer's age<20
         clerk->customer:refuse
         destroy customer
      else customer's age>=20
         clerk->customer:sale
         activate customer
      end
   end
   clerk->cashier:scan bar code
   activate cashier
end
. . .
```

Fig. 5. Difference between combined fragments.

In Fig. 5, the most outside combined fragment in the left has operator, "loop", guard condition, "1,the number of items" and statements of "customer->clerk:items," "activate customer," "activate clerk,", combined fragment "opt,", "clerk->cashier:scan the bar code," and "activate cashier." The most outside fragment in the left has the same statements with the right one. So, we assume that these two combined fragments are the same. Next, we check the inside of the most outside combined fragment.

The one level inside combined fragment in the left has operator, "opt," guard condition, "item=liquor" and statements, "clerk->customer:check the age," and combined fragment, "alt." The one level inside combined fragment in the right has operator, "opt," guard condition, "item=tobacco," and statements, "clerk->customer:check the age," and combined fragment, "alt." In this case, only guard conditions are different with each other, but others are the same, so our tool asks a designer whether these two combined fragments are corresponding to or not.

The most inside combined fragment in the left has the same statements with right one, so we assume that these two combined fragments are the same.

Lastly, summarized report will be generated. An example of a report is shown in Fig. 8.

5   Differential Analyzer

We have developed a prototype system based on our proposed method. This tool is written with Java. The line of source code is about 1700. The procedure of the tool is shown below.

1. This tool reads two files written with PlantUML.
2. If title names are the same in the two files, it adds "old" to the title name of the first file in order to distinguish between two files.
3. It checks status of objects. As for candidates of updated objects, this tool asks a tool user which objects are corresponding to or not.
4. This tool checks status of message passings. As for candidates of updated message passings, this tool asks the tool user.
5. This tool next checks status of execution specifications.
6. It checks status of combined fragments.
7. Based on the status of elements, this tool adds color attributes to original PlantUML statements in order to easily find the differences between two diagrams.
8. Using PlantUML viewer, image files will be generated.
9. This tool shows two image files. (Older diagram will be placed at the left.) and produces a report of differential analysis result.

5.1   Example

Figure 6 shows another cash account system and we compare the original cash account system shown in Fig. 2 with it. The differences between two systems are as below.

1. The sequences of activations of clerk and customer in the combined fragment, loop are different. The difference of the sequence of activations is not semantically different. Our tool does not detect it as difference.
2. The guard conditions in the combined fragment, opt are different.
3. A new combined fragment, opt is added in Fig. 6.
4. Lastly clerk passes thanks message to customer in Fig. 2. This message passing is missing in Fig. 6.

The result is shown in Fig. 7. The tool correctly detected the differences. Blue colored elements in the left diagram in Fig. 7 mean deleted elements in the right. In contrast, blue colored elements in the right diagram mean added elements in the right. Red colored elements mean modified elements. Differential analysis report will be generated as shown in Fig. 8.

5.2   Evaluation

Our tool can correctly detect the following differences between two sequence diagrams as shown in Table 3.

Table 3. Detection of the difference between two sequence diagrams.

Elements	Detected differences
Object	Label, alias, type
Message passing	Label, sender, receiver, sequence
Combined fragment	Label, sequence

Software developers can understand the difference between the two diagrams. A differential analysis report is effective for developers to know the different locations of PlantUML codes.

```
 1) @startUML
 2) title Cash account system for tobacco
 3) actor customer
 4) participant clerk
 5) participant cashier
 6) loop 1,the number of items
 7)    customer->clerk:items
 8)    activate clerk
 9)    activate customer
10)    opt item=tobacco
11)      clerk->customer:check the age
12)      alt customer's age<20
13)        clerk->customer:refuse the sale
14)        destroy customer
15)      else customer's age>=20
16)        clerk->customer:sale
17)      end
18)    end
19)    clerk->cashier:scan the bar code
20)    activate cashier
21) end
22) cashier-->clerk:total amount
23) clerk-->customer:total amount
24) customer->clerk:payment for items
25) clerk->cashier:payment
26) opt total amount >= 10,000 yen
27)    cashier-->clerk:coupon
28)    clerk-->customer:coupon
29) end
30) cashier-->clerk:receipt
31) customer<--clerk:receipt, all the items
32) deactivate clerk
33) deactivate customer
34) @endUML
```

Fig. 6. Cash account system for tobacco.

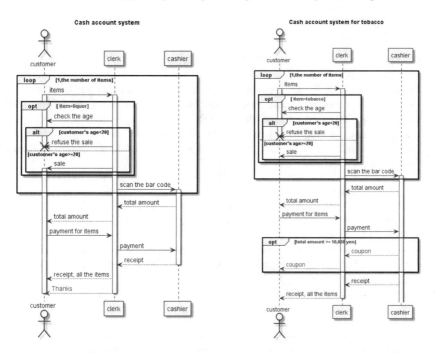

Fig. 7. Differential analysis of sequence diagrams.

Deleted:
  30) message: Thanks from clerk to customer
Modified:
  10) combined fragment: opt item=liquor
  is changed into
  10) combined fragment: opt item=tobacco
Added:
  26) combined fragment: opt total amount => 10,000 yen
  27) message: coupon from cashier to clerk
  28) message: coupon from clerk to customer
  29) end

Fig. 8. Differential analysis report.

6 Conclusion

We have proposed a differential analysis method between two sequence diagrams and developed a prototype system based on the method. Precise evaluation of both the method and the tool is left as a future work.

References

1. Change Vision Inc.: Astah Professional. http://astah.net/editions/professional. Accessed 25 Feb 2020
2. Ekanayake, E.M.N.K., Kodituwakku, S.R.: Consistency checking of UML class and sequence diagrams. In: Proceedings of 8th International Conference on Ubi-Media Computing (UMEDIA), pp. 98–103 (2015)
3. Liu, H., Ma, Z., Zhang, L., Shao, W.: Detecting duplications in sequence diagrams based on suffix trees. In: Proceedings of 13th Asia Pacific Software Engineering Conference (APSEC06), pp. 269–276 IEEE (2006)
4. Matsumoto, A., Yokogawa T., Aman, H.: Consistency verification of UML sequence diagrams modeling sensor networks. In: Proceedings of 8th International Congress on Advanced Applied Informatics (IIAI-AAI), pp. 458-461. IEEE (2019)
5. Mergely: Diff files online. http://mergely.com/. Accessed 25 Feb 2020
6. Mithun, M. Jayaraman, S.: Comparison of sequence diagram from execution against design-time state specification. In: Proceedings of International Conferences on Advances in Computing, Communications and Informatics (ICACCI), pp. 1387–1392. IEEE (2017)
7. PlantUML: Drawing UML with PlantUML, PlantUML Language ReferenceGuide (Version 1.2019.9). http://plantuml.com/en/guide. Accessed 25 Feb 2020
8. Stephan, M., Cordy, J.: A survey of model comparison approaches and applications. In: Proceedings of 1st International Conference on Model-Driven Engineering and Software Development - Volume 1: MODELSWARD, pp. 265–277 (2013)
9. Suyono, H., Nor, K. M., Yusof, S., Rashi, A. H. A.: Use-case and sequence diagram models for developing transient stability software. In: Proceedings of First International Power and Energy Coference (PECon), pp. 109–113. IEEE (2006)
10. Triandini, E., Fauzan, R., Siahaan, D.O., Rochiah, S.: Sequence diagram similarity measurement: a different approach. In: Proceedings of 16th International Joint Conference on Computer Science and Software Engineering (JCSSE), pp. 348–351. IEEE (2019)
11. Wu, S.: The coexistence of sequence diagrams and collaboration diagrams in unified modeling language. In: Proceedings of 4th International Symposium on Knowledge Acquisition and Modeling, pp. 184–187. IEEE (2011)

Model-Based Testing for MQTT Applications

Kotaro Tanabe[1], Yoshinori Tanabe[2,3]($\boxtimes$), and Masami Hagiya[1]

[1] University of Tokyo, Tokyo, Japan
[2] Tsurumi University, Yokohama, Japan
tanabe-y@tsurumi-u.ac.jp
[3] National Institute of Informatics, Tokyo, Japan

Abstract. Model-based testing is a widely-used vital technique for testing software running in a complex environment. In this paper, we propose extensions to existing model-based tools to apply this technique to software that employs the MQ Telemetry Transport (MQTT) protocol for transmitting messages, commonly used in the Internet of Things (IoT) environment. First, in the finite state machine used for generating test cases in a model-based testing framework, we introduce a type of transition that is triggered when receiving MQTT messages. Second, we extend the finite-state machine so that it produces test cases that reflect the characteristics of IoT software – a large number of relatively simple devices communicate with servers. Third, the concept of time is introduced into the finite state machine. Naturally, this is necessary for verifying the properties of software that runs for a long time. Moreover, to facilitate such verification, both real-time and virtual time are introduced. We implemented these extensions into a model-based testing tool, Modbat, and conducted a small experiment to confirm the feasibility, gaining positive results.

1 Introduction

The model-based testing techniques have now been widely accepted as an efficient testing technique. This technique systematically generates test cases from software models, and hence, high-quality test suites that cover corner cases are obtained. The authors applied this technique to cloud software [1] to show that it fits distributed software.

Along with cloud software, IoT software is also attracting attention in the field of distributed systems. In typical IoT systems, small devices, such as sensors participate. The number of devices is usually large, individual devices are not very reliable, and some are prone to failure. The network connecting these devices and the servers is often unreliable.

The aim of this research is to generate test cases for software that runs in such an environment, such as controlling devices, by sending messages to the devices,

---

Supported by JSPS KAKENHI Grant Number 17H01719.

M. Virvou et al. (Eds.): JCKBSE 2020, LAIS 19, pp. 47–59, 2020.
https://doi.org/10.1007/978-3-030-53949-8_5

receiving messages from the devices, analysing the messages, them or managing the device status. More specifically, we concentrate on software systems that use the MQ Telemetry Transport (MQTT) [2] protocol as our system under test (SUT).

The input to such software is dependent on the status of the environment – devices, network, or other factors that affect the input, such as temperature or time of day. Devices for IoT are usually connected to the Internet wirelessly. However, the network communication may become unstable if the devices are far from the router or base station, or if the devices are mobile. In addition, if the size of the battery or antenna of the devices is small, the risk of malfunction due to battery exhaustion or communication failure increases. Owing to the instability of communication and device operation, a number of devices move asynchronously in IoT systems. These features make it challenging to conduct an integrated test of the entire system.

Model-based testing is potentially a good choice to address the testing of such systems. By modelling these factors with transition systems, we can expect various test cases to be generated. This research proposes extensions for model-based testing to support the testing of IoT systems. Our method consists of three approaches.

First, we support MQTT message communication. We propose a method to describe the subscription at the level of the model and enable setting the arrival of a message as a guard condition.

Second, to model the behaviour of a number of devices, we make it possible to share a model (a finite state machine) for many devices. The number of devices operating in IoT systems tends to be large, and the simulation of the individual behaviour would consume enormous computational resources. Therefore, we propose a method to handle them collectively using a transition system. In our approach, the condition of individual devices is treated as a distribution of states in an integrated model. Through this extension, model-based testing scales to large systems. This extension makes model-based testing scalable to an increase in the number of devices.

Third, we introduce the notion of the timeout, which enables us to describe systems that are affected by time. A timeout is a kind of guard condition that is enabled by the passage of time. Using this extension, the model can describe timing properties, such as waiting for a particular input for a limited time. As an implementation of this extension, it is possible to pass real-time given by a transition system. However, this method causes time to wait without doing anything during the test, which makes the test time-consuming. Thus, we propose an implementation that manages virtual time and skips the time passage. We also propose a function to wait real-time for some part of the test because running the whole test of actual systems in virtual time can cause a problem.

We implemented these extensions in model-based testing tool Modbat [3]. In addition, we created test models using these extensions and conducted experiments to examine our contribution to the speedup of test execution time.

The remainder of this paper is organized as follows. Section 2 provides the background, Sect. 3 describes our proposed extensions to Modbat, Sect. 5 presents the results of experiments, and Sect. 6 concludes.

A previous version [4] of this paper was presented, without peer review, at a Japanese domestic workshop.

2   Background

2.1   Modbat

Our study uses the model-based testing tool Modbat [3,5]. This tool uses a model called the extended finite state machine (EFSM) [6] described in a domain-specific language (DSL) based on Scala [7]. EFSM offers the advantage that transition functions are integrated with the runtime environment, and complex data structures or callback functions can be embedded in the model directly [8].

The application of Modbat to real systems includes testing of libraries with non-blocking I/O and exceptions. A previous study [9] found a bug in the Java network library java.nio, and also a race condition in rupy, a high-performance lightweight HTTP server [10]. Another study [1] detected multiple defects in the Apache ZooKeeper, which is a service for maintaining configuration information, naming, and providing distributed synchronization and group services [11].

2.2   MQTT

MQTT is a widely used transport protocol designed for IoT systems. It is supported as one of the principal protocols by many IoT development platforms such as Amazon AWS IoT [12], Microsoft Azure IoT Hub [13], IBM IoT Platform [14], and many more. MQTT (the latest version as of this manuscript is v5.0) is an official OASIS standard [15].

Compared to traditionally used protocols such as HTML, MQTT is lightweight and thus suitable for devices typically used in the IoT environments because the battery consumption is lower.

An MQTT system consists of a server called a broker and clients. It employs a publish/subscribe model. As an example, let us consider a thermometer as an MQTT client. When it reports the current temperature, instead of sending the data to each client who is interested in it, the thermometer "publishes" the data in a "topic name", such as `temperature/japan/tokyo`. The actual data goes to the broker. Other clients who are interested in the data declare that they "subscribe" to the topic beforehand. Thus, the broker can pass the data to the subscribers.

2.3   Related Work

Model-based testing uses abstract models to generate test cases automatically. Typically, concrete test cases are generated from a test model created by a user. There are many model-based testing tools such as Modbat, Spec Explorer [16] and MaTeLo [17].

In this paper, we used the term "model-based testing" in the sense where test cases are generated from finite state machines. Although many model-based

testing tools, such as Modbat, MISTA [18] and MoMuT::UML [19] fall into
this category, any technique that is based on models or that involves models are
generally called model-based testing. For example, models used in FMBT [20] are
written in a language that describes pre-post conditions. Tcases [21] generates
test cases from XML documents that describe functions and input value ranges.

Testing MQTT software is a widely researched topic. Many are interested in
verifying MQTT brokers and libraries for MQTT clients. In [22], a performance
analysis of MQTT is carried out using statistical model checker UPPAAL-SMC.

In [23], a model-based testing technique is applied to MQTT software. The
main target is server software, namely, MQTT brokers. A model was obtained
through the active automata learning method, then it was used to generate
test cases, and they are applied to several MQTT brokers, and found 18 bugs
in total. This research is different from ours as our target is not servers but
clients of MQTT, but it suggests that model-based testing is suitable for testing
MQTT software.

We have previously attempted [24] to incorporate the concept of time into
Modbat. In that research, we introduced the modifier `stay` for transitions, meaning that the next transition is disabled for a specified amount of time, after the
modified transition. It worked in many cases, but was not compatible with our
newly introduced subscription-triggered transition. Therefore, we introduced a
new modifier `timeout` with different semantics in this paper.

3   Extensions to Modbat

In this section, we describe our main contributions to this study. We have
extended Modbat so that:

- sending and receiving MQTT messages can be written in models.
- the following two important concepts can be written in models:
  - Time
  - Number of devices

In the following subsections, we discuss each of these features.

3.1   Subscription-Triggered Transition

Because our target is software that uses MQTT, our model should be able to
handle MQTT messages. Sending messages can be written in methods attached
to transitions in the original Modbat system, without any further help. Receiving
messages is what we need to handle.

We introduce into the Modbat model a type of transition that is triggered
when an MQTT message arrives. In other words, an MQTT message can be
regarded as a precondition for invoking this type of transition. An MQTT topic
should be specified for a transition. Thus, the transition "subscribes" the topic.
Let s be a state of an EFSM and be the source of a transition of this type.
If the current state of an EFSM is s and a message of the topic arrives, then

```
class Class1(...) extends Model { ...
  "A" -> "B" := {  ...
    publish("topic1", "msg 1") ...
  }
  "B" -> "B" := { ...
    publish("topic4", "msg 2"); ...
  } subscribe "topic3"
  "B" -> "A" := { ...
  } subscribe "topic2"
  ... }
```

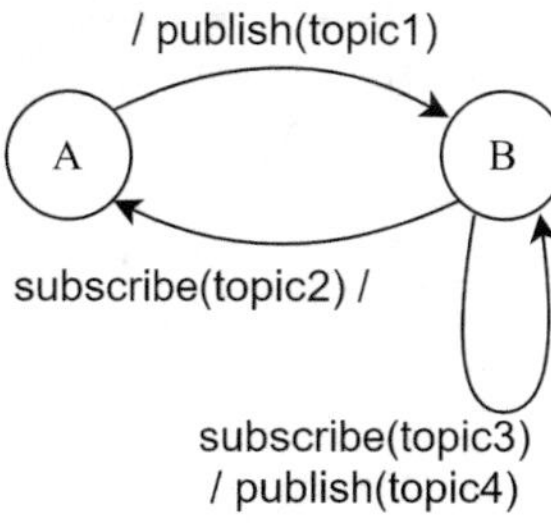

/ publish(topic1)

A   B

subscribe(topic2) /

subscribe(topic3)
/ publish(topic4)

Fig. 1. Subscription-triggerd transition

the transition is invoked, and the message is available in the method attached to the transition. On the other hand, if s is not the current state, the message is ignored (for this EFSM). This behaviour is compatible with the fact that an MQTT client can receive messages only when it subscribes the topic of the messages.

Figure 1 shows a code snippet and a diagram for a model called controller. It has two states A and B. The transition from B to A has a guard `subscribe"topic2"`, meaning that it is only fired when an MQTT message with topic `topic2` is received. The transition from A to B calls `publish` in its attached method, which sends a message to a running MQTT broker.

3.2 State Distribution

In frameworks of model-based testing based on extended finite system machines (EFSM), test cases correspond to paths in the EFSM. To generate test cases on the fly, Modbat keeps the "current state" and moves it along enabled transitions. We can have two or more EFSMs for a system when we have several objects to be considered to generate test cases. In such cases, each EFSM has its current state.

In our application of Modbat for testing MQTT-based software, it is natural to model the behaviour of an IoT device (MQTT clients) with an EFSM. As mentioned above, it is already possible for the original Modbat to have an EFSM for each MQTT client. However, there are several issues to be considered.

First, in IoT environments, the number of MQTT clients may increase. They share an EFSM, but the current states are different. We may be interested in the situation as a whole rather than the state of individual devices. For example, if the EFSM has two states representing normal and failure, we may be interested in the number of devices that are in the failure state. The current Modbat EFSM is not suitable as it assumes independent EFSMs.

Second, in the current implementation of Modbat, a dedicated thread is allocated to each EFSM. Therefore, simulating many devices will result in poor performance.

```
"A" -> "X" := { ... } subscribe(topic1)
"A" -> "Y" := { ... } subscribe(topic2)
"A" -> "Z" := { ... } timeout(10*Const.min)

"B" -> "X" := { ... } timeout(2*Const.hour, 8*Const.hour)

"C" -> "X" := { ... } realTimeout(30*Const.sec)
```

Fig. 2. Transitions with timeout

To address these issues, we have introduced a new type of EFSM. Conceptually, it is a collection of EFSMs that share the same states, transitions and actions. Therefore, it has many current states, or in other words, each state keeps the number of "instances" that stay on the state. The number of instances in each state can be retrieved and used to describe specifications; such as "the ratio of broken devices is less than 3%". Moreover, in order to handle the time-out behaviour (see Sect. 3.3 for details), Modbat tracks the number of instances that enter the state in each time slice.

3.3   Timeout

Many systems behave depending on time. Some sensor systems may send their reports periodically, for example, once in an hour. Some control systems may wait for a message from devices that it monitors for a certain period of time, and if no message comes in, it may judge that the device is out of order.

To describe these types of system behaviour, we introduced the concept of timeout into the Modbat model.

Formally, timeout is a property of transitions, with the amount of time. For example, in Fig. 2, a ten-minute timeou t is attached to the transition from A to Z. The transition is fired 10 min after an instance enters A if no message with topic *topic1* or *topic2* is arrived during the period. Another type of timeout has two parameters, such as the one from B to X. In this case, the expiration time is chosen randomly between 2 h and 8 h.

One state can have at most one transition with timeout attached, so there are no races between transitions.

3.4   Virtual and Real Timeout

Waiting for all timeouts in real time would not be realistic. We cannot wait for four hours when we conduct a system test. An obvious workaround would be to reduce time by a fixed percentage, but it is not ideal for the following reasons.

First, even with a reduced rate, we still suffer from unnecessary waiting time, which will not become zero. Second, some timeouts absolutely require the specified amount of time. For example, we have MQTT message exchanges between our running models, and it takes some time for a message to be delivered.

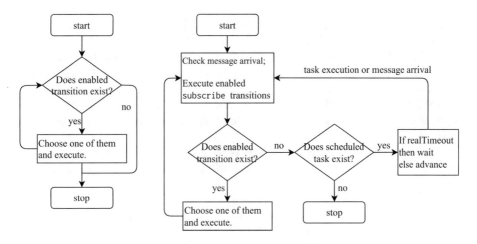

Fig. 3. Changes in the main loop of Modbat

Therefore, if we try to decide whether a device is in failure state by sending a message and receiving a response, we need to wait for a specific time, and it cannot be reduced for speed-up.

Therefore, we have introduced two different types of timeout: real and virtual. A real timeout expires only when the specified amount of time elapses physically. On the other hand, when a virtual timeout is stored, we record the expiration time for the timeout. When the system enters a status where no other transition can be (immediately) invoked, then all the pending virtual timeouts are checked, and the current time is "skipped" to the earliest expiration time.

We also have EFSMs with more than one instance, as described in Sect. 3.2. Instances that share the same current state may have different expiration times. Therefore, simply counting the number of instances in a state is not sufficient. However, if we kept expiration time for each instance independently, the advantage of collecting instances would be lost. To address this issue, we internally define a value *timeSlice* and split time with this interval. Instances that fall into the same interval are regarded as a group, and a representative timeout expiration time is recorded for the group.

4   Implementation

The introduction of three extensions had a major impact on the main loop of Modbat. The original main loop, shown in the left-hand side of Fig. 3, is rather simple: it chooses one of the enabled transitions randomly and executes it, while there are such transitions.

We modified the main loop is as follows: It first chooses a state with enabled transitions (without timeout or subscription), if there is some. If two or more enabled transitions exist, it divides instances staying on the state according to the weight (probability) of the transitions, and execute them. Once, it becomes a situation where there are no enabled transitions, Modbat checks whether scheduled tasks (transitions with timeout whose source state has waiting instances) exist. If all scheduled tasks are in virtual time, Modbat advances the virtual time to the earliest expiration time. If there is one with a real timeout, Modbat sleeps until the earliest timeout expiration time (including virtual time). If a message arrives while sleeping, Modbat wakes up and fires transition that subscribes the topic of the message. The right-hand side of Fig. 3 illustrates the modified main loop.

A task is scheduled when a model instance enters a state, which does not have enabled transitions but has a transition with a timeout. If the timeout has only one expiration time, it is simply recorded with the number of model instances. If the timeout has two parameters, the model instances are evenly divided into groups according to the value of *timeSlice*, which is configurable by the user, and the expiration time for each group is recorded.

The virtual timeout is implemented using an Akka mock scheduler [25]. This scheduler provides basic features for handling virtual time, such as skipping a fixed amount of time. We implemented the necessary features for Modbat, such as advancing the virtual time spontaneously, on top of this scheduler.

5   Evaluation

To demonstrate the effectiveness of our methods, we conducted a small experiment. First, we designed a small MQTT application and built the corresponding Modbat models to confirm that the features introduced in this paper are sufficient to describe the testing environment. Second, we ran the model to see the following:

- whether the virtual timeout works as expected so that the runtime is shorter than the case of real time,
- whether the multi-instance EFSM works as expected so that it reduces the runtime compared to many single-instance EFSMs.

These experiments test a system of smart wattmeters. This system consists of meters and a controller communicating through MQTT. The meters measure electric power in all rooms in a house. The controller receives the data, sums them, and reports the total electrical power to the user. There is a probability that some meters may be broken. When the controller detects some broken meters, it sends an alarm to the operator of the system.

In these experiments, we tested the behaviour of the controller. We verified that the controller reports the correct value and that it provides a warning if and only if some meters are broken.

5.1   Models

A model of the smart house consists of five types of model instances, as shown in Figs. 1, 2 and 4. Meter has multiple instances, while the other four models have only one instance each.

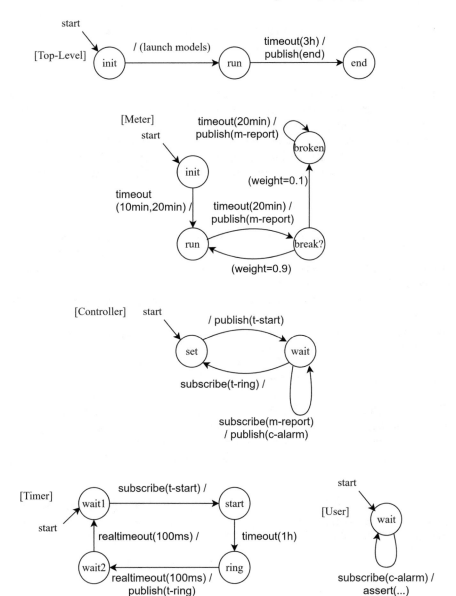

Fig. 4. Models used in the experiments

Top-Level. The instance of this model is created at the start of the simulation. It launches instances of other models in the transition from state *init* to state *run*. After waiting for some virtual time in *run*, it sends an MQTT message that notifies the end of the test session to a topic called **end**. All the states in other models have a transition subscribing to this topic, which leads an instance to the final state of the model. (Related transitions and states are omitted in the figures.)

Meter. A meter is installed in each room. At first, all meters are running correctly. Each meter measures the electric power in the room, and reports it to the controller every twenty minutes, using topic m-report. After the transition, each meter returns back to the *run* state with probability 90% and goes to the *broken* state with probability 10%, which means the meter is broken. Meter instances in the *broken* state report watt to the controller repeatedly like correctly-running meters, but they report an incorrect value. If a meter is broken, it will not be repaired during the simulation.

Controller. The controller receives messages from meters, sum them up, and reports the result to the user every hour. First, the controller sends a message to the timer using topic t-start. Then, it waits until the timer notifies the passage of one hour using a message with topic t-ring. While waiting, the controller receives values from the meters through messages with topic m-report and sums them up. In addition, if the received value is much higher or much lower than the average of previously received values, the controller assumes that the meter that sent the irregular value is broken, and publishes an alarm message to the user using topic c-alarm. After a message from the timer arrives, the controller reports the summed value to the user.

Timer. The timer starts when it receives a message with topic t-start. After waiting for one hour by virtual time, it publishes a message that tells the passage of one hour with topic t-ring. Considering the time of MQTT communications, the timer waits for a brief duration of real-time before and after publishing the message.

User. This is a model of a tester of the controller. The user subscribes topics m-report and c-alarm. It asserts that the value reported to m-report is correct and that the controller publishes a message to c-alarm if and only if some meters are broken.

5.2   Experiments and Results

We conducted two experiments to demonstrate the effectiveness of the methods described in the previous chapter. The experiments show the speedup of the test execution time by introducing virtual time and the shared model. We performed experiments on Ubuntu 16.04 on Oracle Virtual Box. The host OS is Windows 10 Pro, running on a machine with Intel Core i7-6600U CPU and 8GB memory.

   To evaluate the effect of virtual time on test time, we run the tests of the models described above with two variants. One uses keyword timeout for the timeout of the transition from state run to end in model *Top-Level* (from now on referred to as transition end-trans), which means the test is executed in virtual time. The other uses keyword `realTimeout` for transition `end-trans`, which means the test is executed in real time. We have converted the value of timeout of `end-trans` and have run tests for each value.

The results are shown on the left-hand side of Table 1. Execution time is the average of five runs. The execution time of the code with `timeout` is shortened, while that of the code with `realTimeout` is approximately the same as the timeout of `end-trans`.

Table 1. The results of experiments

Amount	Execution time (sec)		# of devices	Execution time (sec)	
	Virtual	Real		Indiv.	Shared
1 min	3.9	62.2	10	3.1	3.3
10 min	3.5	602.2	100	20.0	13.3
			200	37.0	18.4
			1000	137.4	36.4

We also conducted an experiment to evaluate the effect of introducing models with two or more instances. In this experiment, we run the tests of the same models as the first experiment with two types of source code. The two codes are different in how to handle the model of devices. One operates the instances of the meters individually, and the other operates a distribution of the meters.

The results are shown on the right-hand side of Table 1. As the number of devices increases, the execution time is faster when managing by distribution in comparison to running instances individually.

6   Conclusion

We proposed extensions for model-based testing for IoT systems and implemented them in a model-based testing tool Modbat. The first extension makes it possible for transitions to be fired when an MQTT message is received. By the second extension, we can have models that reflect the status of many instances. Two types of timeout features, real and virtual, are introduced as the third extension. We have built a small experimental test environment that uses the extensions and confirms that it works as expected functionally and shows performance improvements compared to the original version.

Future tasks include providing a virtual time library for SUTs. Currently, if an SUT issues the sleep method, then it naturally sleeps real-time. Instead, if the SUT uses the library, virtual time is applied when running under Modbat while keeping the original behaviour (sleep) on production.

Another direction is implementing dedicating MQTT broker for Modbat. This will allow the user, when testing the MQTT broker is out of the scope, to remove real-time timeout for communication, and enable Modbat to run completely in virtual time. In addition, this extension will make it easier to simulate an unstable network by controlling packet loss or delay of delivery in Modbat.

Acknowledgements. The authors would like to thank Professor Cyrille Artho for many helpful discussions and valuable comments.

References

1. Artho, C., Gros, Q., Rousset, G., Banzai, K., Ma, L., Kitamura, T., Hagiya, M., Tanabe, Y., Yamamoto, M.: Model-based API testing of Apache ZooKeeper. In: International Conference on Software Testing, Verification and Validation (ICST 2017), pp. 288–298 (2017)
2. Rahul, G., Andrew, B.: MQTT version 3.1.1. http://docs.oasis-open.org/mqtt/mqtt/v3.1.1/os/mqtt-v3.1.1-os.html
3. Artho, C., Biere, A., Hagiya, M., Platon, E., Seidl, M., Tanabe, Y., Yamamoto, M.: Modbat: a model-based API tester for event-driven systems. In: Hardware and Software: Verification and Testing, pp. 112–128 (2013)
4. Tanabe, K., Tanabe, Y., Hagiya, M.: Speedup of model-based testing for IoT software using virtual time and state distribution of devices. IEICE Tech. Rep. **119**(392), 37–42 (2020)
5. Artho, C., Biere, A.: Modbat. https://people.kth.se/~artho/modbat/
6. Cheng, K.T., Krishnakumar, A.S.: Automatic functional test generation using the extended finite state machine model. In: Design Automation Conference, DAC 1993, pp. 86–91 (1993)
7. Odersky, M., Spoon, L., Venners, B.: Programming in Scala. Artima, Mountain View (2008)
8. Artho, C., Havelund, K., Kumar, R., Yamagata, Y.: Domain-specific languages with Scala. In: Formal Methods and Software Engineering, pp. 1–16 (2015)
9. Artho, C., Hagiya, M., Potter, R., Tanabe, Y., Weitl, F., Yamamoto, M.: Software model checking for distributed systems with selector-based, non-blocking communication. In: International Conference on Automated Software Engineering (ASE), pp. 169–179 (2013)
10. Larue, M., Martino, E., Funk, M., Chen, A., Lee, A., Lung, C., Hoyt, D., Baghdasaryan, H.: rupy - a tiny Java nio HTTP application server (2013)
11. Hunt, P., Konar, M., Grid, Y., Junqueira, F., Reed, B.: Zookeeper: wait-free coordination for internet-scale systems. In: ATC. USENIX, vol. 8 (2010). Yahoo Research
12. Barr, J.: AWS IoT – cloud services for connected devices. https://aws.amazon.com/blogs/aws/aws-iot-cloud-services-for-connected-devices/
13. Communicate with your iot hub using the MQTT protocol. https://docs.microsoft.com/en-us/azure/iot-hub/iot-hub-mqtt-support
14. Getting to know MQTT. https://developer.ibm.com/articles/iot-mqtt-why-good-for-iot/
15. MQTT v5.0 OASIS standard. https://docs.oasis-open.org/mqtt/mqtt/v5.0/mqtt-v5.0.html
16. Veanes, M., Campbell, C., Grieskamp, W., Schulte, W., Tillmann, N., Nachmanson, L.: Model-Based Testing of Object-Oriented Reactive Systems with Spec Explorer, pp. 39–76 (2008)

17. Model based testing tool - discover MaTeLo — ALL4TEC. https://www.all4tec.com/
18. Xu, D., Thomas, L., Kent, M., Mouelhi, T., Traon, Y.L.: A model-based approach to automated testing of access control policies. In: Proceedings of ACM symposium on Access Control Models and Technologies (SACMAT 2012), pp. 209–218 (2012)
19. Krenn, W., Schlick, R., Tiran, S., Aichernig, B., Jobstl, E., Brandl, H.: MoMut::UML model-based mutation testing for UML. In: Proceedings of International Conference on Software Testing, Verification and Validation (ICST), pp. 1–8 (2015)
20. FMBT. https://01.org/fmbt/
21. Tcases. https://github.com/Cornutum/tcases
22. Houimli, M., Kahloul, L., Benaoun, S.: Formal specification, verification and evaluation of the MQTT protocol in the Internet of Things. In: International Conference on Mathematics and Information Technology (ICMIT), pp. 214–221 (2017)
23. Tappler, M., Aichernig, B.K., Bloem, R.: Model-based testing IoT communication via active automata learning. In: Proceedings of International Conference on Software Testing, Verification and Validation (ICST), pp. 276–287 (2017)
24. Yoneyama, J.: Model-based testing simulating unstable networks and devices for IoT software. Master's thesis, University of Tokyo (2018)
25. Github - miguno/akka-mock-scheduler: a mock akka scheduler to simplify testing scheduler-dependent code. https://github.com/miguno/akka-mock-scheduler

A Study of Mechanism Design for C2C Service Based on Multi-agent Simulation

Takuya Izumisawa$^{(\boxtimes)}$, Yuki Katsumata, and Akira Yamada

Research Laboratories, NTT Docomo, Inc., 3-6 Hikari-no-oka, Yokosuka-shi,
Kanagawa 239-8536, Japan
takuya.izumisawa.ns@nttdocomo.com

Abstract. In Consumer-to-Consumer (C2C) services where individuals provide their own idle assets to other individuals, the number of trouble incidents between individuals is increasing. These incidents arise because individuals act inappropriately (defection strategy) when providing or using assets against the will of the counterpart. One goal for C2C services is to activate the market by increasing the number of individuals who take appropriate action (cooperation strategy). Toward this end, we propose a mechanism that achieves the desired cooperation rate. The number of individuals who follow the cooperation strategy will increase as incentives are increased, and there is a trade-off between the achievable cooperation rate and incentives. The purpose of this study is to clarify the conditions that achieve the desired cooperation rate with fewer incentives. Simulation results show that the proposed mechanism increases the number of individuals who follow the cooperation strategy and that incentives contribute to achieving the desired cooperation rate rather than penalties.

Keywords: Evolutionary game · Mechanism design · Multi-agent simulation

1 Introduction

The Sharing Economy is an economic activity that makes idle assets owned by an individual available to other individuals. In Sharing Economy services (Consumer-to-Consumer (C2C) services), transactions are conducted between individuals who provide their own idle assets (providers) and individuals who use those idle assets (consumers). Here, idle assets include not only tangible assets such as living space, goods, and money, but also intangible assets such as a person's time and skills. In many C2C service models, providers and consumers are matched through the Internet, transactions between individuals are performed by giving and receiving money, and a part of the received money is allocated to C2C operators as fees.

In recent years, through the proliferation of smartphones and development of social networking services it has become easy to match providers to consumers, and the C2C service market is expanding worldwide [1]. However, with its expansion, the number of trouble instances between individuals that use C2C services is increasing rapidly. These incidents are caused by both providers and consumers, and arise because individuals act

© Springer Nature Switzerland AG 2020
M. Virvou et al. (Eds.): JCKBSE 2020, LAIS 19, pp. 60–69, 2020.
https://doi.org/10.1007/978-3-030-53949-8_6

inappropriately (defection strategy) when providing or using assets against the will or desire of the counterpart. Therefore, the occurrence of these incidents is the main factor that hinders further increase in the number of C2C service users. In order to activate the market, we need to increase the number of individuals who take appropriate action (cooperation strategy).

In order to deter trouble between individuals, a mutual evaluation system has been introduced. In the system, individuals mutually evaluate correspondence in the transaction whenever one is conducted. Thereby, individuals can predict whether the counterpart will follow the cooperation strategy before the transaction and can take steps to avoid individuals with a low evaluation.

However, trouble in C2C services still exists because some individuals do not place importance on their own evaluation. Therefore, in order to increase the number of individuals who follow the cooperation strategy, it is necessary to design a mechanism in addition to the mutual evaluation system.

Toward this end, we propose a mechanism that changes the number of incentives or penalties based on evaluation points. It is self-evident that the number of individuals who follow the cooperation strategy will increase as the incentives are increased. However, there is an upper limit to the incentives that can be given to individuals, and thus there is a trade-off between the achievable cooperation rate and incentives. The purpose of this study is to clarify the conditions that achieve the desired cooperation rate with fewer incentives in the proposed mechanism.

In this paper, we implement the proposed mechanism based on a multi-agent simulation formulated from evolutionary game theory. Furthermore, we confirm the change in the cooperation rate according to given incentives and penalties. In this simulation, agents repeat transactions and learn which strategy, either cooperation or defection, should be followed to maximize their gain. We evaluate the cooperation rate according to the given parameters. Based on this evaluation, we clarify the gain conditions required to achieve the desired cooperation rate.

In Sect. 2, we overview evolutionary game theory and related research. In Sect. 3, we describe the proposed mechanism and implemented simulation model. In Sect. 4, the evaluation environment and results are presented. We give our conclusions in Sect. 5.

2   Mechanism Design Based on Evolutionary Game

2.1   Evolutionary Game Theory

Evolutionary game theory is a methodology that introduces dynamics to game theory and explores the final consequences of a game. This methodology was introduced to evolutionary biology to analyze the evolution of living organisms [2]. However, it is currently being actively studied and applied to various fields such as social sciences and economics.

One important issue in evolutionary game theory is a mechanism design to generate cooperative behavior by individuals in a certain social environment. In regard to this mechanism design, Nowak et al. [3] discuss five reciprocal mechanisms that have the potential to promote and maintain cooperative behavior by individuals. One of them, network reciprocity, is a mechanism that generates cooperation in the entire network by limiting the network that connects individuals to having only the cooperation strategy or the defection strategy. In this mechanism, individuals only transact with specific individuals (neighbors) on the network and imitate strategies from those neighbors. Network reciprocity has attracted attention as a mechanism used to reproduce real-world social networks such as relationships between individuals [4].

2.2 Related Research

Zhang *et al.* [5] and Xie *et al.* [6] introduced a mechanism based on individual evaluations in C2C services called crowdsourcing. Crowdsourcing is a mechanism that allows a person to order or accept work freely by providing skills possessed by an individual to other individuals via a platform on the Internet. The types of work that can be ordered include system development, illustration creation, and video shooting. However, crowdsourcing faces the free-rider problem [7], i.e., individuals do not provide results according to a counterpart's request. To address this problem, Zhang *et al.* designed a mechanism in which individuals with high evaluations earn more rewards. Their results showed that individuals satisfy the requests of the counterparts in order to obtain more incentives. Xie *et al.* designed a mechanism to penalize individuals with a low evaluation. Their results showed that their mechanism eliminates individuals who provide results that do not satisfy the counterpart requests.

In addition, Yang *et al.* [8] and Luo *et al.* [9] introduced a mechanism based on individual evaluations in C2C services called cloud-sensing. Cloud-sensing is a mechanism that accumulates a large amount of data from sensors built into mobile devices such as smartphones and wearable-devices owned by individuals, and estimates real-world aspects from the data [10]. This has attracted attention because it can be applied to a wide range of fields such as environment monitoring, traffic condition monitoring, healthcare, and opinion polling. In cloud-sensing, which encourages the provisioning of large amounts of sensor data, there is a problem in motivating people to provide sensor data. To address this, Yang *et al.* designed a mechanism in which individuals who supply higher quality data gain more incentives. Their results showed that individuals can be motivated to provide high-quality data through their mechanism. Luo *et al.* designed a mechanism in which a quality threshold is set for the provided data, and a higher reward is given to individuals when the data supplied from an individual are above the threshold, and vice versa. Their results showed the superiority of this mechanism.

These related studies suggest that individuals take appropriate actions according to set incentives and penalties based on the evaluation of that individual. We hypothesize that C2C services, which include intangible assets such as a person's time and skills, and tangible assets such as living space, goods, and money would perform similarly.

3 Simulation Model

3.1 Game Definition

In the employed model, we define a game that references the simulation model by Chica *et al.* [11]. The agent strategy in the C2C service sets two discrete values for each provider and consumer. Each agent selects one of the four strategies given below. The total number of agents is equal to the total number of agents in each strategy.

- Trusted Provider (TP): a provider who follows the cooperation strategy
- Untrusted Provider (UP): a provider who follows the defection strategy
- Trusted Consumer (TC): a consumer who follows the cooperation strategy
- Untrusted Consumer (UC): a consumer who follows the defection strategy

Table 1 shows the game payoff structure in the C2C service. Temptation T is the gain by the consumer (UC) over the provider (TP). Reward R is obtained from the transaction between the provider (TP) and the consumer (TC). Sucker S is the loss gained by the provider (TP) when the consumer strategy is UC, and X represents the profit gained by the provider (UP) regardless of the consumer strategy.

Table 1. Game payoff structure in C2C service.

		Consumers	
		TC	UC
Providers	TP	R, R	$-S, T$
	UP	$X, -X$	$X, -X$

In the payoff structure shown in Table 1, the relationship between each payoff is as follows.

$$2R > T > R > S > X \geq 0 \qquad (1)$$

In this simulation, we set $T = 1$, $S = 0.5$, and $X = 0.25$. Here, R is the reward exchanged between individuals in the conventional C2C service, and is a parameter of reward that can be freely set by the C2C service operator. In this study, we hypothesize that C2C service operators can control the transactions between individuals in the market and achieve the desired cooperation rate by designing a mechanism in which R is determined based on the evaluation points given to individuals. Specifically, this is described in Sect. 3.3.

3.2 Proposed Mechanism Design

In this paper, we propose a mechanism that increases the cooperation rate for individuals in C2C services. The proposed mechanism applies incentives or penalties based on mutual evaluation. Specifically, individuals receiving a high evaluation receive incentives according to the mutual evaluation point difference in addition to rewards they should receive, and vice versa.

In this mechanism, reward R_i is obtained by individual i after the completion of a transaction between individual i and counterpart j as follows.

$$R_i = \begin{cases} R + \alpha_{Gain}\left(\frac{|Eva_i - Eva_j|}{Eva_{max} - Eva_{min}}\right), & Eva_i \geq Eva_j \\ R + \alpha_{Loss}\left(\frac{|Eva_i - Eva_j|}{Eva_{max} - Eva_{min}}\right), & Eva_i < Eva_j \end{cases} \qquad (2)$$

At this time, α_{Gain} and α_{Loss} are the incentive and penalty weighting factors, and Eva_i and Eva_j are the reputation points of individual i and the counterpart j, respectively. This Eva is not the evaluation point in each transaction, but the average point of the evaluation points accumulated up to these transactions. Terms Eva_{max} and Eva_{min} are the maximum and minimum evaluation points, respectively. In this simulation, we use a five-level evaluation method, so $Eva_{max} = 5$ and $Eva_{min} = 1$. In this mechanism, the payoff calculated as R in Table 1 is replaced with R_i based on the evaluation score of each individual, and the payoff is calculated.

3.3 Multi-agent Simulation

We implement the proposed design based on the multi-agent simulation formulated from evolutionary game theory. The details regarding the simulation flow are given hereafter. (i) First, agent i selects either the cooperation strategy or defection strategy. (ii) Agent i randomly selects a counterpart from neighbors linked in the same network and transacts as a provider or a consumer. At this time, if both the agent and the selected counterpart select roles as providers or consumers, the transaction is not conducted. In addition, the transaction execution rate is determined according to the evaluation point of the counterpart. (iii) After the transaction, agent i obtains the payoffs given in Table 1 and (1) based on the strategies of itself, the trading partner, and their evaluation points. (iv) After obtaining the payoff, agent i is evaluated based on the strategy followed in step (ii) by the counterparts. (v) Finally, agent i selects either the cooperation strategy or defection strategy to follow in the next transaction. Then, steps (ii) to (v) are repeated.

4 Experiments and Results

4.1 Experimental Configuration

In this experiment, the network model adopts the Scale-Free model based on the Barabási-Albert (BA) algorithm [12]. The number of vertices is set to 100, which is equal to the total number of agents. Furthermore, the average network degree is set to eight. Each agent repeats the evolution game on these networks.

In step (ii) in Sect. 3.3, we set a selection bias regarding whether or not agent i should transact based on the evaluation points of agent j (counterpart). Specifically, the more evaluation points that agent j has, the greater the increase in the probability that agent i executes the transaction, and vice versa. Figure 1 shows the transaction execution rate that agent i will execute a transaction with agent j. In this experiment, we compare two scenarios for the cooperation rate considering each parameter. In the first scenario, individuals trade regardless of the counterpart evaluation (Fig. 1, Bias 0). In the other scenario, the number of trading opportunities decreases for individuals with a low evaluation. Specifically, in this scenario (Fig. 1, Bias 1–4), the transaction execution rate changes stepwise with evaluation point 3 as the threshold. We assume four selection bias settings with different transaction execution rates in each pattern.

In step (iv) in Sect. 3.3, each agent receives an evaluation point with the probability according to the agent strategy given in Table 2.

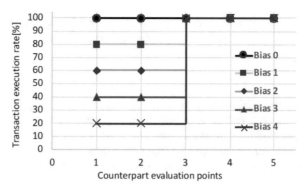

Fig. 1. Agent transaction execution rate.

Table 2. Relationship between followed strategy and probability of being given an evaluation point.

		Evaluation point				
		5	4	3	2	1
Strategy	Cooperation	40%	30%	20%	10%	0%
	Defection	0%	10%	20%	30%	40%

In step (v) in Sect. 3.3, each agent decides the strategy to follow in the next transaction based on the Pairwise-Fermi rule. This rule is a model in which an individual chooses a random neighbor existing in the same network and decides whether or not to imitate actually the neighbor strategy according to the payoff difference between itself and the neighbor. The probability that agent i imitates the strategy of agent j is given below.

$$P_{i \leftarrow j} = \frac{1}{1 + \exp\left(\frac{\Pi_i - \Pi_j}{\kappa}\right)} \qquad (3)$$

Here, Π is the total payoff obtained by an agent and κ is set to 0.1. This equation shows that the probability that agent i imitates the neighbor agent strategy increases as the total payoff becomes lower than the total payoff of the neighbor agent.

In step (i) in Sect. 3.3, we initialize the agent coordination rate at 50% and set the evaluation points of all agents to 3. All agents perform steps (ii) to (v) as one-time steps, and continue until the difference between the cooperation rate at the end of a certain time step and the cooperation rate of the previous time step becomes sufficiently small, or this one-time step is repeated 3000 times. If the difference is sufficiently small, the cooperation rate is used as the value at the end of the simulation. If 3000 time steps are performed, the cooperation rate is used as the average of the cooperation rate in the last 100 time steps.

In this experiment, the cooperation rate is shown when the value of R is changed from 0.6 to 0.9 in steps of 0.1 and the values of α_{Gain} and α_{Loss} are changed from 0 to

1 in steps of 0.1 as parameters. The simulation is iterated 100 times, and the ensemble average of the cooperation rate for each parameter is calculated.

4.2 Results

First, we confirm the total number of agents for each evaluation point in the case of the proposed mechanism that gives an incentive or penalty according to the evaluation point, and in the case of the conventional mechanism that does not provide such. Figure 2 shows the total of 100 iterations for the number of agents for each evaluation point at the end of the simulation. In the figure, there is no selection bias in either set of results. The proposed mechanism shows the results when R is set to 0.9 and α_{Gain} and α_{Loss} are set to 1. On the other hand, the conventional mechanism shows the results when R is set to 0.9 and α_{Gain} and α_{Loss} are set to 0. Focusing on both mechanisms, the average evaluation point is 3.16 in the former and 2.43 in the latter. These results suggest that the proposed mechanism increases the number of individuals who follow the cooperation strategy compared to that for the conventional mechanism. Based on this, we consider that individuals in C2C services follow the cooperation strategy and are amenable to the will of the counterpart to gain incentives or avoid penalties.

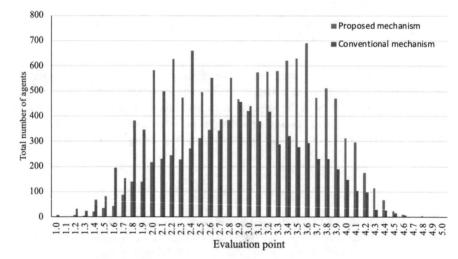

Fig. 2. Total number of agents for each evaluation point.

Next, we confirm the change in the cooperation rate due to the difference in each scenario. Figures 3 and 4 show the ensemble average of the cooperation rate calculated for each parameter $(r, \alpha_{Gain}, \alpha_{Loss})$. Figure 3 shows the cooperation rate when there is no selection bias (Bias 0), and Fig. 4 shows the cooperation rate when the selection bias is most prominent (Bias 4). The blue region represents where the cooperation rate is 50% or more (cooperation region), and the red region represents where the cooperation rate is 50% or less (defection region). Comparing the case without selection bias (Fig. 3) to that with selection bias (Fig. 4), the latter case reduces the number of trading opportunities for individuals with a low evaluation and mitigates the conditions to achieve the desired cooperation rate. Based on this, we consider that controlling opportunities to trade is a factor in increasing the number of individuals that assume a cooperation strategy. This is because an individual with a low evaluation cannot increase his/her payoff if fewer trading opportunities are available. Therefore, the number of individuals who follow the cooperation strategy can be increased at a low cost by introducing a mechanism that decreases the number of opportunities for individuals with a low evaluation.

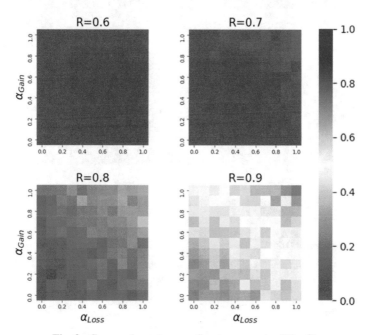

Fig. 3. Cooperation rate according to parameter (Bias 0).

Finally, we confirm the conditions of each parameter necessary to satisfy the desired cooperation rate. Based on the above results, the condition of each parameter that satisfies the desired cooperation rate is predicted by linear regression analysis. This time, we assume the desired cooperation rate is 50%. Figure 5 shows the parameters that achieve a cooperative rate of 50% for each bias condition. Focusing on the slope of the graph, the slope of the straight line is greater than -1 for all bias conditions except for bias condition 3. This shows that the value of α_{Gain} affects the increase in the cooperation rate more than the value of α_{Loss}. These results suggest that individuals in C2C services will assume a cooperation strategy with strong motivation to acquire incentives rather than to be penalized.

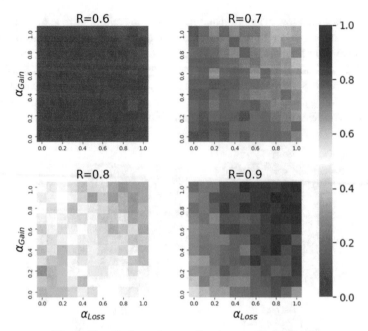

Fig. 4. Cooperation rate according to parameter (Bias 4).

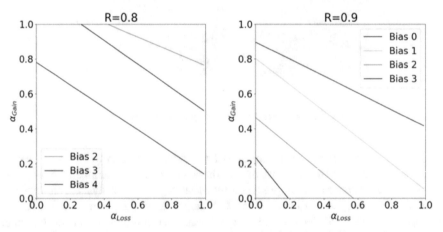

Fig. 5. Conditions for each parameter that satisfies 50% coordination rate.

5 Conclusions

One goal for C2C services is to activate the market by increasing the number of individuals who take appropriate action (cooperation strategy). Toward this end, we propose a mechanism that achieves the desired cooperation rate. The number of individuals who follow the cooperation strategy will increase as incentives are increased, and there is a trade-off between the achievable cooperation rate and incentives. The purpose of this study is to clarify the conditions that achieve the desired cooperation rate with fewer incentives. In this paper, in order to clarify the conditions, we implemented the proposed mechanism based on the multi-agent simulation formulated from evolutionary game theory and confirmed the change in the cooperation rate according to given incentives and penalties. The simulation results show that introducing the proposed mechanism into the C2C service market with appropriate gain parameters may increase the number of individuals assuming a cooperation strategy. By evaluating the coordination rate for each gain parameter, we confirmed that individuals will assume a cooperation strategy with strong motivation to acquire incentives rather than to be penalized. Therefore, the number of individuals who follow the cooperation strategy can be increased at a low cost by introducing a mechanism with parameters set to emphasize incentives or a mechanism that decreases the number of opportunities for individuals with a low evaluation.

References

1. PwC: The sharing economy-sizing the revenue opportunity. PwC UK Website (2016)
2. Smith, J.M., Price, G.R.: The logic of animal conflict. Nature **246**, 15–18 (1973)
3. Nowak, M.: Five rules for the evolution of cooperation. Science **314**, 1560–1563 (2016)
4. Nowak, M.: Evolutionary games and spatial chaos. Nature **359**, 826–829 (1992)
5. Zhang, Y.: Reputation-based incentive protocols in crowdsourcing applications. In: IEEE INFOCOM (2012)
6. Xie, H.: Incentive mechanism and protocol design for crowdsourcing systems. In: Proceedings of IEEE Allerton (2014)
7. Feldman, M.: Free-riding and whitewashing in peer-to-peer systems. IEEE J-SAC **24**, 1010–1019 (2004)
8. Luo, T.: Sustainable incentives for mobile crowdsensing: auctions, lotteries, and trust and reputation systems. IEEE ComSoc (2017)
9. Yang, S.: On designing data quality-aware truth estimation and surplus sharing method for mobile crowdsensing. IEEE J-SAC **35**, 832–847 (2017)
10. Raghu, K.: G: Mobile crowdsensing: current state and future challenges. IEEE Commun. Mag. **49**, 32–39 (2011)
11. Chica, M.: An evolutionary trust game for the sharing economy. In: IEEE CEC (2017)
12. Barabási, A.: Emergence of scaling in random networks. Science **286**, 509–512 (1999)

Experimental Evaluation of Traceability Checking Tool for Goal Dependency Modeling

Haruhiko Kaiya[1]([⊠]), Wataru Fujita[1], Ryotaro Yamada[1], Atsuo Hazeyama[2], Shinpei Ogata[3], Takao Okubo[4], Nobukazu Yoshioka[5], and Hironori Washizaki[6]

[1] Kanagawa University, Hiratsuka 259-1293, Japan
kaiya@acm.org
[2] Tokyo Gakugei University, Tokyo 84-8501, Japan
[3] Shinshu University, Nagano 380-0928, Japan
[4] IISEC, Yokohama 221-0835, Japan
[5] NII, Tokyo 100-0003, Japan
[6] Waseda University, Tokyo 169-8555, Japan

Abstract. In a complex socio-technical system, a human's goal is delegated to many actors such as human and machines. Because the delegated goal can be decomposed into several sub-goals by each actor, goals are delegated recursively until an actor provides the means to achieve each sub-goal. We have already proposed a notation and a method called GDMA to represent and analyze the issues above. Because GDMA can be represented in a class diagram, software engineers do not have to use specific tools of GDMA models. To confirm whether a goal is properly achieved by suitable means, we have to trace such delegation and decomposition relationships. However, it is not easy to confirm it in a real-world system because of the system's complexity. In this paper, we present a tool to check such traceability. The tool is implemented as a plugin of an existing UML modeling editor, and goal dependencies and decompositions are depicted using color. We also evaluate the tool through a comparative experiment. As a result, the tool enables an analyst to check the traceability without omission although it does not improve efficiency of the traceability checking task.

Keywords: Goal oriented requirements analysis · Istar · Traceability · Modeling tool · Experimental evaluation

1 Introduction

In recent years, a lot of intelligent systems are introduced into practice. Such introduction sometimes makes us regard there are few or no human roles and efforts in our daily activities. For example, a system recommends some goods suitable for each customer when he/she is shopping around. Prediction of the fluctuations in the market helps or navigates traders. However, human has to

© Springer Nature Switzerland AG 2020
M. Virvou et al. (Eds.): JCKBSE 2020, LAIS 19, pp. 70–83, 2020.
https://doi.org/10.1007/978-3-030-53949-8_7

inherently play many important roles including some decisions and tasks because most activities are social systems. Therefore, we have to regard each daily activity as a Socio-Technical System (STS).

Modeling a STS is helpful to explore effective introduction of systems. Goal-oriented requirements analysis (GORA) is one of the promising ways in this exploration because of the following reasons. First, GORA makes us identify human goals to be achieved. Second, GORA makes us specify how to decompose a goal into other goals. Third, GORA makes us clarify who achieves a goal after all. To explore appropriate system introduction, an analyst has to trace dependencies and decomposition of goals without omission. In this paper, we focus on traceability of such dependencies and decomposition. Managing such traceability contributes to at least the following two issues. First, analysts and stakeholders can understand how a goal is decomposed into sub-goals recursively and how each of them is achieved after all. Second, they can understand why the goal should be achieved. We call the former issue as forward traceability check, and the latter one as backward traceability check.

Istar (usually written as i*) [14] is a famous GORA, and many variations of istar exist. We also proposed a variation [7] called Goal Dependency Model with Attributes (GDMA), and developed its modeling tool [6]. An analyst of GDMA has to manually follow traceability of goal decomposition and dependencies as same as analysts of other istar variations do. However, the analyst cannot analyze the model well because manual and subjective checking of traceability is not comprehensive. Although the representation of GDMA is complex as well as istar, the algorithm of tracing them can be defined. However, it is not easy for an analyst to follow the algorithm without omission. We thus developed a tool to trace such decomposition and dependencies of GDMA. We also evaluated the tool through a comparative experiment. We report the tool and its evaluation in this paper.

The rest of this paper is organized as follows. In the next section, we explain a modeling language GDMA, which is a variation of istar. We also introduce its modeling tool. As well as istar, GDMA is suitable to describe models of STSs. A tool for checking traceability on a GDMA model is then presented in Sect. 3. To evaluate the usefulness of the tool, we performed an experiment. The experiment is reported in Sect. 4. Section 5 briefly reviews related works, and finally we summarize our current results and show our future issues.

2 Modeling Language GDMA and Its Modeling Tool

2.1 Modeling Language

We proposed a graphical modelling language GDMA [7] to address the following research questions (RQs).

1. How can requirements analysts encourage stakeholders to introduce machines into their STS? We call artificial systems as machines according to Jackson's concept [5].

2. How can analysts demonstrate that such introduction will be beneficial to all stakeholders?

Although the second RQ is not related to the traceability checking tool in this paper, we also explain the characteristic of GDMA related to this RQ, i.e. eighth characteristic below, so that GDMA is comprehensively understood.

GDMA is a variation of istar [14]. To introduce new machines into an STS written in a GDMA model, we systematically analyzed the model. The nine main characteristics of GDMA are below.

1. An individual, an organization, or an artificial element such as software applications and hardware who wants or achieves a goal is called *an actor* in the same way as that in istar. When an actor is an artificial element, we call it an artificial-actor or *a machine*. Otherwise, we call an actor *a human-actor*. We place a stereo-type «*machine*» in the graphical notation in GDMA, when an actor is a machine.
2. A goal is specified by a noun phrase and a past perfective verb such as "a reservation for my hotel room made".
3. *Strategic Dependency (SD)* between actors in istar is used for specifying the delegation of goal achievement. *A dependee* in an SD is an actor who will be able to achieve a goal, and an actor who wants to achieve a goal in an SD is called *a depender*. Even if a dependee is identical to a depender, i.e., his/her goal is achieved by him/herself, the SD is explicitly written.
4. And/or decomposition is specified in each actor in the similar way as the *Strategic Rationale (SR) Model* in istar.
5. *Means-end* in istar is taken into account, but the means is not specified in detail. We simply specify that an actor achieves a goal in his/her SR model by him/herself. A means-end in our graphical notation is depicted in a hexagon icon.
6. Goals delegated to an actor are rigorously managed. A goal should be delegated to an actor, if and only if the goal is neither means-ended nor decomposed in an SR model.
7. Soft-goals or quality requirements are not first class objects but attributes of a goal. As a result, a goal in our model is a functional goal.
8. Numerical values are attached to a depender and each dependee for each attribute of a goal in an SD. A value to a depender called the *want-level* indicates to what extent the depender wants the goal to be achieved. A value to a dependee called the *can-level* indicates to what extent the dependee can achieve the goal.
9. Tasks and resources are not used.

By using values in the eighth characteristic above, we defined metrics so that we can examine whether a modeled STS is preferable with respect to human. We do not explain the metrics in detail because they are out of scope of this paper. When we introduce some machines into the STS on the model, we have to decide whether the machines are preferable or not. The metrics contribute to such decision.

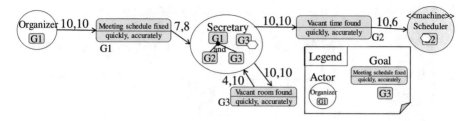

Fig. 1. Example of a GDMA model

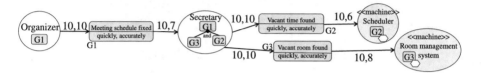

Fig. 2. Another GDMA model after introducing additional machine into the model in Fig. 1

We will explain GDMA by using the example of its model in Fig. 1. A classic problem of a "meeting scheduler" [14] is partially modelled in this example. The three actors of an "Organizer", "Secretary", and "Scheduler" are depicted as circles, and the SRs of each actor are depicted in each circle.

Goals are depicted by the rounded rectangle and each goal has three types of attributes: an ID, a label, and quality attributes. "G1", "G2", and "G3" in the figure are IDs. The labels indicate the content of the goals such as "Meeting schedule fixed" or "Vacant room found". The quality requirements of individual goals are specified as quality attributes. For example, two quality attributes "quickly" and "accurately" are attached to each goal.

Three SDs are depicted in the figure: the first is between the "Organizer" and "Secretary" via "G1", the second is between the "Secretary" and "Scheduler" via "G2", and the last is between "Secretary" and "Secretary" via "G3". If a goal has the same ID as another, these two goals are identical. For example, "G2" in the SR model of "Secretary" is the same as goal "G2" in the second SD. The label and quality attributes of a goal can be omitted if the same goal is depicted elsewhere and the goal has the label and quality attributes.

Want levels of 10 and 10 are attached to "Organizer" and 7 and 8 can-levels are attached to "Secretary" in the first SD. These values mean that "Organizer" wants to achieve goal "G1" extremely (10) quickly and accurately, but "Secretary" can achieve the goal almost (7) quickly and very (8) accurately. The value takes 1 (weakly wanted or achieved) to 10 (strongly wanted or achieved). These values enable us to specify the balance between a depender's wants and a dependee's ability. There is an imbalance in the attribute "quickly" because the want-level of "Organizer" is 10 but the can-level of "Secretary" is 7 in the first SD. The intuitive meaning of the values in the SD is as follows: "Although an organizer wants to fix meeting schedule extremely quickly, his secretary cannot fix it so quickly."

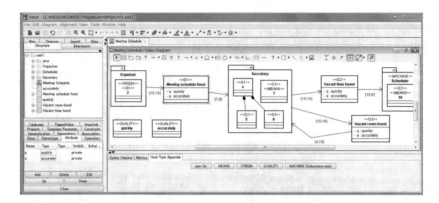

Fig. 3. Whole GUI of the modeling tool

When "G1" is delegated to "Secretary", she decomposed it into two sub-goals "G2" and "G3". She then delegates "G2" to a system "Scheduler", and achieves "G3" by herself. Even when a depender achieves his/her own goal by his/herself, the goal is once delegated as shown in Fig. 1. Two occurrences of "G3" therefore appear in SR of "Secretary" in the figure.

2.2 Modeling Tool

As shown in Fig. 1 and 2, the notation of GDMA is not easy to be managed in the same way as istar. We thus defined UML based notation of GDMA, and we also developed a modeling tool [6] on the basis of existing modeling tool called astah [1]. The modeling tool is developed as a plugin of the astah modeling tool. The plugin mechanism enables us to access model elements in real time on the basis of the even based architecture. Figure 3 shows a snapshot of the tool, where a model in Fig. 1 is edited. The tool provides the following functions: syntax checker, metrics calculation and some input support, i.e. putting specific stereo-types easily. If its user has already used astah modeling tool, he/she can use our GDMA modeling tool without hesitation.

3   Traceability Checking Tool for GDMA

Our modeling tool in Sect. 2.2 is helpful when its user describes a model and examine whether machines introduced in an as-is model are useful or not. However, the user has to manually follow the traceability of goal delegation and decomposition. As mentioned in introduction, the user cannot confirm whether a goal is delegated to suitable actors without such traceability. In addition, he/she cannot confirm whether a goal is properly decomposed without it.

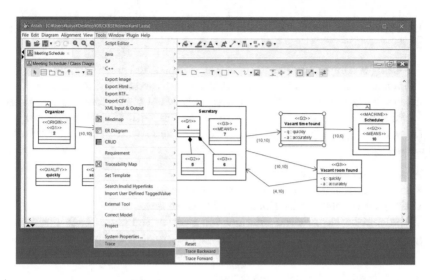

Fig. 4. Choose an originate goal and a command in the Traceability Checking Tool

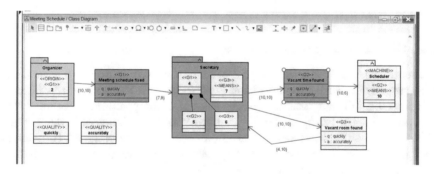

Fig. 5. The result of Traceability Checking Tool

To help such a user to follow the traceability without omission, we developed a traceability checking tool as shown in Fig. 4 and 5. The tool is also developed as an astah plugin in the same way as our modeling tool. The user of the tool in the figures wants to know why G2 "Vacant time found" should be achieved. The user thus performs the backward traceability function. The usage of the tool is as follows.

1. A user chooses a goal, which is a source or destination of a trace. In Fig. 4, G2 is chosen.
2. He/she chooses a command of "Trace Forward" or "Trace Backward" from a menu. In Fig. 4, a trace backward command is chosen.
3. The tool shows the trace with colors. Intermediate actors and goals become red, and terminal actors become green. Figure 5 shows the result of a backward trace. The user can know a goal G1 exists due to an actor "Organizer".

He/she can also know "Secretary" and goal G1 are intermediate actors and goals.

By choosing a command "Reset", the user restores the colors on the tool.

This traceability checking tool is developed by extending the modeling tool mentioned above. Traceability is checked only when the syntax check is passed. Therefore, we don't have to take care of models with the wrong syntax such as the loop of goal delegation. The outline of the forward trace algorithm is as follows.

1. A source goal is chosen.
2. An actor who is delegated the goal is identified.
3. If the goal is achieved by the actor, the trace ends.
4. If the goal is decomposed into several goals in the actor, each decomposed goal becomes new source goal and we go back to step 1 above. Note that both and/or decompositions are processed in the same way.

The algorithm of backward trace is almost the reverse.

4   Evaluation

4.1   Objective of the Experiment

As mentioned in introduction, a goal of human is decomposed into several sub-goals recursively. Such decomposition is performed on the basis of the rationale of each actor. Without tracing such a series of decomposition, there are following two problems. First, we cannot confirm whether a sub-goal is delegated or assigned to a suitable actor. For example, a sub-goal is assigned to a human actor even though it can be achieved systematically and automatically. Second, we cannot find unnecessary sub-goals. Although all sub-goals should be parts of human's goals, we cannot confirm it without traceability of goal decomposition.

We assume it is not easy for analysts without any support to trace such delegation and decomposition comprehensively and fast when the model is large. We also assume our tool helps the analysts to do so. We thus confirm the assumptions through a comparative experiment mentioned below.

4.2   Objects, Metrics and Subjects

To confirm the assumptions above, the model used in the experiment should be large enough. If the model is small, most analysts without the tool can check the traceability among it comprehensively and fast. We thus prepared a model of a socio-technical system on the basis of a famous case study "London Ambulance Services" (LAS) [2]. The model contains 10 actors and 14 goals between the actors. One of the authors described the model.

We prepared two types of questions to confirm the assumptions. We asked each subject to answer the questions. The first type is about forward traceability among goal dependency. The example is as follows:

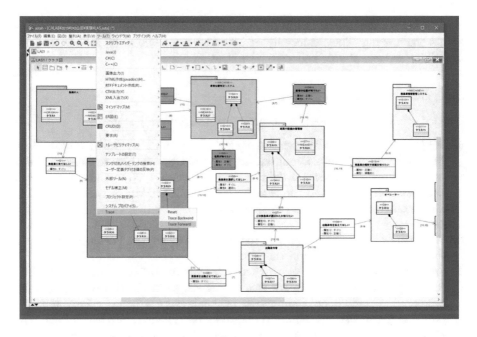

Fig. 6. Typical usage of this tool in this experiment

A patient wants to achieve a goal "an ambulance comes to her". Please enumerate the goals that finally contribute to achieving the goal. In addition, please enumerate actors who related to this achievement.

The second type is about backward traceability. The example is as follows:

A navigation system wants to achieve a goal "an address of the patient known". Please indicate a goal which originally causes the goal. In addition, please enumerate actors who related to this achievement.

To confirm the assumptions mentioned in Sect. 4.1, we define the following two metrics:

$$ACC = \frac{\text{The number of right goals and actors answered by a subject}}{\text{The number of right goals and actors}} \times 100$$

TIME = Minutes during traceability checking by a subject.

The ACC metric corresponds to recall rather than precision because we regard no missing links are important. With the tool, we expect ACC becomes larger than ever, and TIME becomes smaller than ever. Because our tool always tells each subject right goals and actors, ACC of tool users is usually 100%. The subjects learnt it in advance. We did not ask subjects obey the results of the tool. We just told them they might use the tool in the treatments with the tool.

We asked 8 bachelor course students as subjects in this experiment. They were 4th year bachelor students. Although they did not have industrial experiences, they had studied usual software engineering lectures including requirements engineering. They especially had studied istar, KAOS [9] and GDMA by describing models about their own problems.

4.3   Steps of the Experiment

The experiment was performed according to the following steps.

1. We gave a lecture and an exercise of GDMA and its traceability tool in 21 of November, 2019. It took about one hour. We also asked them to consent to the usage of their experimental data in technical papers. Note that all subjects had a similar lecture and the exercise only about GDMA in November 2018. The lecture and the exercise in this step was a kind of review of the GDMA notation.
2. We briefly review GDMA and its tool in 28 of November. We then performed the experiment as follows in the same day.
3. We first asked each subject to check traceability by using the traceability checking tool. Typical usage of this tool is shown in Fig. 6. Because all subjects are Japanese people, the model elements were written in Japanese. We provided four questions, half of them were first type question and others were second type. The questions were documented and printed on papers, each of which was distributed to each subject.
4. Each subject answered the questions. The answers were submitted to a learning management system (LMS) on the Web. The LMS was familiar to all subjects because it was frequently used in their lectures. The spending time was measured respectively.
5. We second asked each subject to check traceability manually, i.e. without the checking tool. Note that they could use the modeling tool. We provided four questions in the same way as step 3 above.
6. Each subject answered the questions in the same way as step 4 above.
7. We asked each subject to answer questionnaires so that we could know qualitative evaluation of the tool by the subjects.

Because our tool just show the results of traceability, it inherently does not contribute to learning how to follow the traceability. Therefore, we regard the step 3 never improves the ability how to follow the traceability on GDMA before the step 5, i.e. manual traceability checking. However, the step 3 will improve the understanding of LAS activities in the model before the step 5.

The list of questionnaires is as follows.

1. Does the tool make you check traceability faster than ever?
2. Does the tool contribute to improving the accuracy of traceability checking?
3. What else the tool is useful for you?
4. Is the tool easy to use?
5. What kinds of additional functionalities are wanted?
6. Please let us know the problems about its usability.
7. Please let us know its bugs.

Table 1. Results of the experiment

Subject ID	With tool		Without tool	
	ACC	TIME	ACC	TIME
a	100.0	28	74.1	22
b	100.0	31	96.3	28
c	100.0	32	29.6	26
d	100.0	33	96.3	35
e	100.0	34	74.1	29
f	100.0	28	88.9	25
g	100.0	41	92.6	45
h	100.0	41	44.4	37
Average	100.0	34	74.5	31

4.4   Results and Discussion

Table 1 shows the results of this experiment. The first row shows the ID of subjects. The second and third rows show ACC and TIME in step 4, i.e. with the tool. The fourth and fifth rows show ACC and TIME in step 6, i.e. without the tool. The lines except the last one correspond to the results of each subject. The last line shows the average of each metric.

We first focus on the difference of ACC with or without the tool, i.e. the second and fourth rows in the table. As shown in the table, they seem to be different. By using t-test, p value is about 0.024. Therefore, we may say ACC without the tool is significantly different from ACC with the tool. According to the second questionnaire, all subjects wrote the tool contributed to improving the accuracy. We cannot identify the reasons why ACC without tool is not good. We can assume several reasons, for example, complexity of the model and/or misunderstanding of questions and the model. Another experiment should be performed to clarify such reasons.

We second focus on the difference of TIME with or without the tool, i.e. third and fifth rows in the table. By using t-test, p value is about 0.085. Therefore, we cannot say TIME without the tool is significantly different from TIME with the tool. According to the first questionnaire, all subjects wrote the tool made them check traceability faster than ever.

According to the fourth questionnaire, seven subjects wrote the tool was easy to use. The main reasons are the visualization and coloring features of this tool. One subject wrote the tool was not so easy to use because the tool did not highlight the start point of traceability. According to the sixth questionnaire, we got several proposals from subjects. For example, more precise coloring was required. Because functionalities of the tool can be called only from the menu, some subjects wrote it was laborious. A feature such as a short-cut for calling the functionalities was thus required. No bugs were reported through the questionnaire.

4.5　Threats to the Validity

Threats to internal validity can affect the independent variables of an experiment. In this experiment, only a single group was used. In addition, all treatments were performed in the same day. Therefore, the threats to internal validity exist. Because the same model used in two subsequent treatments, i.e. with and without the tool, the second treatment without tool had some advantages. This could be regarded as a threat to the internal validity. However, the results of second treatment were not better than the results of first one. If this threat can be removed, we can show using tool is more preferable than the current results show. If we ask different subjects for each treatment, we can remove the threat.

Threats to external validity reduce the generalizability of the results. Unfortunately, we cannot say our subjects were typical requirements analysts. We thus say there was this type of threat. In addition, we used only one model. This also causes the threats to the external validity. According to Table 1, subjects c and h seem to be the outlier because ACC without tool is too bad. We do not regard them as the outlier because analysts without sufficient knowledge about the model can exist in real world.

Construct validity is the degree to which the variables measure the concepts they are to measure. Our metrics ACC and TIME seem to be suitable for measure what we want to know. ACC does not take into account the answers based on incorrect traceability links. We regard the fact will not badly affect the actual analysis of the traceability because such incorrect answers can be rejected during the review of traceability links. If some results of our tool is not right, ACC with the tool does not become 100%. If some subjects did not obey the tool results, ACC with the tool could be less than 100%. As far as we examined the results, the tool seems to give us the right answers, and the subjects obeyed the results. We can suspect subjects with the tool simply obeyed its results, and the subjects did not understand what they had to do. According to the results in Table 1, we suspect two of them, subject c and h, seemed not to do. However, other six subjects seemed to understand it because ACC without tool was not so bad.

Conclusion validity is concerned with the relationship between the treatment and the outcome. Because we achieved a statistical test, we believe the results were reliable.

5　Related Work

According to our previous survey of software traceability techniques [8], there were a few studies of using traceability. Especially, the quality of its usage was rarely studied. Instead, the quality of traceability recovery was frequently studied in many studies. In this research, we experimentally analyzed the effects on using traceability by a supporting tool. Accuracy and efficiency of its usage were analyzed in this paper.

Because GDMA is a variation of istar, we discussed traceability usage and its tools in istar and its variations. We think traceability contributes to understanding and comprehending an STS in a model. In many studies, rigorous reasoning, transformation and pure visualization are used for such understanding. In [12], visualization of istar is used to understand actor's behavior. However, how to follow the visual model seems to be up to an analyst. In [13], model elements were prioritized and modularized so that analysts can easily analyze and maintain the model. Most of them were just the proposal of techniques. In [10], comprehension of contribution links is discussed. This is out of scope in our research because our GDMA does not have contribution links.

Finally, we review recent tools of istar. We can find tools for analyzing models and for transforming them. In [3], a tool transforming use cases into istar was proposed. However, we cannot find tools directly for tracing traceability. Istar 2.0 editor on Eclipse was proposed in [11], but its analytic functions were not clear. Creative leaf [4] was a tool to support creativity in requirements engineering. Therefore, it did not help us to follow traceability based on delegation and decomposition.

6   Conclusion

We developed and experimentally evaluated a tool for following the traceability on Goal Dependency Model with Attributes (GDMA). Because GDMA is a variation of istar, each goal in a model is delegated to an actor such as human or a system. The goal is then decomposed into several sub-goals if it is not achieved by the actor. Without following the traceability caused by such delegation and decomposition, an analyst cannot confirm whether a goal is achieved by suitable means. In addition, he/she cannot easily understand why a goal is delegated. Through a comparative experiment, our tool contributed to checking the traceability without omission. However, the tool did not contribute to improving the efficiency in checking the traceability. Because each subject reviewed a model to answer some questions with our tool first, and then he/she reviewed it to answer others without the tool, the treatment with the tool is in an adverse condition. We thus conclude our tool is useful for checking the traceability.

According to the results of questionnaires, we want to improve the user interface in our tool. Such improvement will give a positive influence on the efficiency of using our tool. A specific feature of GDMA is attaching numerical attributes on each goal dependency. The current traceability checking tool does not mind this feature. Without checking these numerical attributes, some problems can occur in a model. For example, suppose a goal should be achieved "so accurately" (the attribute about 10 is given). Although decomposed and delegated sub-goals of the goal should basically inherit this quality attribute "so accurately", there is no means to check it. We want to extend our traceability checking tool to provide the means.

Acknowledgment. This work was supported by JSPS KAKENHI Grant Numbers 18K11249 and 16H02804.

References

1. Change Vision, Inc.: Astah professional. https://astah.net/products/astah-professional/. Accessed May 2020
2. Finkelstein, A., Dowell, J.: A comedy of errors: the London ambulance service case study. In: IWSSD 1996, p. 2. IEEE Computer Society (1996)
3. Geraldino, G.C.L., Santander, V.F.A.: The JGOOSE tool. In: Pimentel, J., Carvallo, J.P., López, L. (eds.) Proceedings of the 12th International i* Workshop co-located with 38th International Conference on Conceptual Modeling (ER 2019), Salvador, Brazil, 4 November 2019. CEUR Workshop Proceedings, vol. 2490. CEUR-WS.org (2019). http://ceur-ws.org/Vol-2490/paper14.pdf
4. Horkoff, J., Maiden, N.A.M.: Creative leaf: a creative iStar modeling tool. In: Proceedings of the Ninth International i* Workshop co-located with 24th International Conference on Requirements Engineering (RE 2016), Beijing, China, 12–13 September 2016, pp. 25–30 (2016). http://ceur-ws.org/Vol-1674/iStar16_pp25-30.pdf
5. Jackson, M.: Problem Frames, Analyzing and Structuring Software Development Problems. Addison-Wesley, Boston (2000)
6. Kaiya, H., Haga, K.: A CASE tool for goal dependency model with attributes based on an existing UML editor. In: Knowledge-Based and Intelligent Information & Engineering Systems: Proceedings of the 21st International Conference KES-2017, Marseille, France, 6–8 September 2017, pp. 1196–1205 (2017). https://doi.org/10.1016/j.procs.2017.08.033
7. Kaiya, H., Ogata, S., Hayashi, S., Saeki, M.: Early requirements analysis for a socio-technical system based on goal dependencies. In: New Trends in Software Methodologies, Tools and Techniques - Proceedings of the Fifteenth SoMeT_16, Larnaca, Cyprus, 12–14 September 2016, pp. 125–138 (2016). https://doi.org/10.3233/978-1-61499-674-3-125
8. Kaiya, H., Sato, R., Hazeyama, A., Ogata, S., Okubo, T., Tanaka, T., Yoshioka, N., Washizaki, H.: Preliminary systematic literature review of software and systems traceability. In: Knowledge-Based and Intelligent Information & Engineering Systems: Proceedings of the 21st International Conference KES-2017, Marseille, France, 6–8 September 2017, pp. 1141–1150 (2017). https://doi.org/10.1016/j.procs.2017.08.152
9. van Lamsweerde, A.: Requirements Engineering: From System Goals to UML Models to Software Specifications. Wiley, Hoboken (2009)
10. Liaskos, S., Alothman, N., Ronse, A., Tambosi, W.: On the meaning and use of contribution links. In: Proceedings of the 12th International i* Workshop co-located with 38th International Conference on Conceptual Modeling (ER 2019), Salvador, Brazil, 4 November 2019 (2019). http://ceur-ws.org/Vol-2490/paper5.pdf
11. Liebenberg, M., Roßmaier, K., Lakemeyer, G.: An iStar 2.0 editor based on the eclipse modelling framework. In: Proceedings of the 10th International i* Workshop co-located with the 29th International Conference on Advanced Information Systems Engineering (CAiSE 2017), Essen, Germany, June 12–13 2017, pp. 55–60 (2017). http://ceur-ws.org/Vol-1829/iStar17_paper_18.pdf
12. Pant, V., Yu, E.S.K.: Using i* to reason about employee behavior on public social media. In: Proceedings of the 12th International i* Workshop co-located with 38th International Conference on Conceptual Modeling (ER 2019), Salvador, Brazil, 4 November 2019 (2019). http://ceur-ws.org/Vol-2490/paper4.pdf

13. Tapia, L.M., López, L., Ayala, C.P.: Dealing with goal models complexity using topological metrics and algorithms. In: Ghanavati, S., Liu, L., López, L. (eds.) Proceedings of the 10th International i* Workshop co-located with the 29th International Conference on Advanced Information Systems Engineering (CAiSE 2017), Essen, Germany, 12–13 June 2017. CEUR Workshop Proceedings, vol. 1829, pp. 43–48. CEUR-WS.org (2017). http://ceur-ws.org/Vol-1829/iStar17_paper_3.pdf
14. Yu, E., Giorgini, P., Maiden, N., Mylopoulos, J.: Social Modeling for Requirements Engineering. The MIT Press, Cambridge (2010)

Adoptability Assessment
of AI Service Systems

Hironori Takeuchi$^{(\boxtimes)}$, Yukino Oyama, and Kazuki Yamamoto

Musashi University, Tokyo, Japan
h.takeuchi@cc.musashi.ac.jp, {s1182037,s1182163}@st.musashi.ac.jp

Abstract. In this study, we consider the adoption of artificial intelligence (AI) service systems developed in a business domain to other domains. We propose a method to assess whether we can apply an existing AI service system to a different business task. We identify conditions for the adoption of the AI service system, and through experiments, confirm that we can predict whether the example inputs for a new business task can work for the existing system by referring to the conditions without the support of data scientists.

Keywords: AI service system · Adoptability · Enterprise architecture · OODA process · Knowledge intensive data-processing system

1 Introduction

There are several machine-learning-based artificial intelligence (AI) programming modules that are available as application programming interfaces (APIs). Therefore, it is now possible to utilize AI technologies for practical business applications. Currently, enterprises are applying these technologies for supporting inquiry service operators to answer queries about business operations, products, or services, and managing screening operations that use documents containing different types of client data. As a result, there are numerous AI business applications released in various industries. In companies, decision makers such as business executives learn success cases of AI solutions that are already applied in other industries and consider that they can introduce such AI solutions easily into their own business and improve their business performance.

In this paper, we consider AI system development projects under such a situation. Generally, projects encounter numerous points of contention and technical challenges when AI service systems are being developed [1]. It is reported that there are nine reasons for the failure of machine-learning projects[1]. Among those reasons, the following are related to project management.

---

[1] https://www.kdnuggets.com/2018/07/why-machine-learning-project-fail.html.

© Springer Nature Switzerland AG 2020
M. Virvou et al. (Eds.): JCKBSE 2020, LAIS 19, pp. 84–95, 2020.
https://doi.org/10.1007/978-3-030-53949-8_8

1. Asking the wrong questions
2. Trying to use AI to solve the wrong problems
3. Not having suitable data

When we introduce an existing solution using AI technologies in a domain into our own business, these problems can be solved during the project's planning phase by assessing both the solution and the AI service system developed in it. This assessment can be conducted effectively if data scientists collaborate with project members from the business division. However, in many cases, data scientists cannot be expected to be involved in the project planning stage, because of limited human resources [8]. Therefore, although project members from the IT division should lead the assessment without supports of data scientists, their expertise is not explicitly externalized.

In this study, we consider a method for assessing the adoptability of the existing AI service systems used in enterprise functions. We introduce a model representing an existing AI service system used in enterprise functions and derive service-system-level conditions that should be satisfied when it is introduced into a different business task. In an experiment using an AI speaker, we show that we can predict whether an existing AI service system can work for example inputs in the target business domain through our method, and confirm that our method can be used for the adoptability assessment by project members without the support of data scientists.

The remainder of this paper is structured as follows. In Sect. 2, related works are described. In Sect. 3, we introduce the AI service system and define the research hypothesis of this study. In Sect. 4, we describe the model and method for the adoptability assessment. In Sect. 5, we present an experiment to test how the proposed method is applied and confirm the effectiveness of the method. Finally, we discuss the experimental results in Sect. 6, and summarize the key points and future work in Sect. 7.

2 Related Work

To apply a machine-learning engine that is one of the main AI technologies trained by the data in one domain to a different domain, a method called transfer learning can be used [9]. In this method, the trained model is modified based on the difference between the various domain data. In this study, we assess whether a solution using AI technologies developed in one domain can be applied to another domain. Therefore, we need to represent the solution at the system level and assess its adoptability. Projects that implement big data analytics or machine-learning technologies require a representation of the system architecture in which the solution is to operate [4]. Some system architectures have been proposed [3,6]. However, the domain-specific information or features of the business where the system is used are not represented. Therefore, we cannot use these architectures directly for the adoptability assessment.

Business–IT alignment is introduced in enterprise system management. In organizations, business–IT alignment defines the processes and objectives by which business and IT functions are to be integrated. This approach decreases organizational uncertainty and improves enterprise agility. Methods for constructing a business–IT alignment model using an enterprise architecture (EA) approach were introduced in [7,10]. For the AI service system development project, an EA-based business–IT alignment model was proposed [12]. The business–IT alignment model can be used to assess whether we can apply an existing AI service system of a business task into our own business task. However, although we need conditions or limitations on the existing AI service system for the assessment, such information is not represented in business–IT alignment models.

Research has been conducted on a generic process for developing a system to implement machine-learning technologies [1], and the roles of data scientists in projects [8] are discussed. In this study, although the applicability assessment of machine-learning technologies in the target business domain or adoptability assessment of the existing machine-learning model is regarded as a process during the model-selection phase [1], practical assessment methods are not discussed.

3   AI Service System and Research Hypothesis

3.1   Enterprise Architecture and AI Service System

In this study, we present an AI service system by employing an EA modeling approach, using ArchiMate [13] as the EA modeling language. With ArchiMate, a business and IT system can be represented as an EA by the three layers: business, application, and technology. In a practical AI system development project, AI technologies can be used as an API running on a cloud environment. Therefore, we focus on the business and application layers.

Figure 1 shows a general business application represented using ArchiMate.

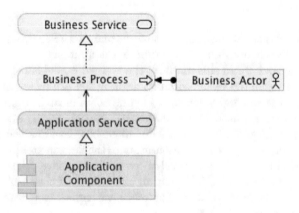

Fig. 1. General business application represented using ArchiMate

In our model, to represent AI service systems, we employ the following three business concepts and three application concepts:

- Business service: an explicitly defined and executed business activity.
- Business process: a sequence of business activities that produce a planned outcome.
- Business object: a set of concepts used within a particular business domain.
- Application service: an explicitly defined and exhibited application behavior.
- Application component: an element of application functionality aligned to the implementation structure.
- Data object: data structured for automated processing.

Next, we define the AI service systems considered in this study. There are three types of human intelligence: analytical intelligence, creative intelligence, and practical intelligence [11]. In our study, we consider developing a system with analytical intelligence for business functions that support human activities or are their substitutes. Analytical intelligence selects and outputs the optimal one from a predefined set of options for the given input data [11]. Machine-learning technologies are used when we replicate this intelligence in a software system. To use machine learning for system development, we define options for the target business domain and collect example inputs related to each option. A machine-learning model is generated (trained) from a training dataset containing such pairs of options and examples. This model is deployed into a runtime machine-learning engine, which obtains input data and makes output data available using the predictive model. We assume that AI service systems contain one or more machine-learning engines for realizing analytical intelligence.

The enterprise functions considered in this study are divided into business tasks, to each of which practitioners are assigned. We consider that an AI service system containing machine-learning engines supports these business tasks. Each business task can be seen a decision-making process. The OODA (Observing, Orienting, Deciding, Acting) model is proposed by Boyd as a basis for the decision-making process. This model illustrates that human decision-making processes comprise the following four steps [2]:

- Observing the facts by capturing relevant data about the environment.
- Orienting the human participants to the information derived from the facts by applying knowledge.
- Deciding on the directives based on the operational hypothesis.
- Acting on the directives and producing outcomes.

A business task is described using the OODA model and ArchiMate, as shown in Fig. 2. In this model, each step is defined as a business process and the inputs and outputs of each step are represented as business objects. Their definitions are shown in Table 1. The practitioners explicitly read, write, or refer to these business objects when executing each step. In this study, we assume that AI service systems support this type of business task.

3.2   Research Hypothesis

In this study, we consider a situation where we introduce an existing AI service system developed for a different enterprise function to our own enterprise functions.

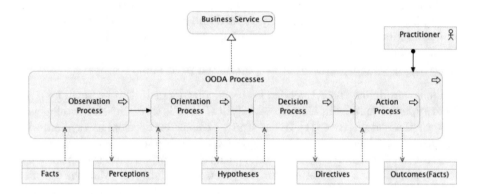

Fig. 2. Business task represented by the OODA model

Table 1. Business objects in the OODA model

Business Object	Definition
Facts	Inputs of the business task
Perceptions	Extracted and converted data just required for the business task
Hypotheses	Structured information for the objective of the business task
Directives	Candidates of the options for the action
Outcomes	Outputs of the business task (inputs for the next business task)

Nearest every day, business solutions using AI service systems and their business values are described in business news articles, industry reports, etc. In companies, decision makers such as business executives in the business division consider that they can introduce such AI service systems easily and improve their business effectively when reading such articles or reports. However, in real cases, we can rarely introduce a business solution using an AI service system without any customization development. In some cases, it is found that the development cost required for adoption in the trial project is significant, and as a result, business executives become disappointed with the AI technologies.

To introduce an existing business solution using an AI service system to a different enterprise function, project members from the business and IT divisions must assess the following questions without supports of data scientists in the early stages of the project-planning phase.

– Can typical use cases in the new enterprise function work for the AI service system?
– When a use case does not work, which conditions in the AI service system should be satisfied?

In the machine-learning technology, there is a technique called transfer learning, in which we adopt the machine-learning model trained by the data in one domain to another domain [9]. In this study, we consider the domain adoption of AI service systems comprising some machine-learning engines. In this situation, we need to identify the system-level conditions that the system should be satisfied with for the adoption. We propose a method to assess the adoptability of AI service systems by responding to the above questions. If we can derive conditions where an AI service system can be introduced into a new domain, we can predict whether the service system can work for the example inputs in the new domain without supports of data scientists and assess whether we should introduce the existing AI service system or design a new one from scratch. We confirm this use case through an experiment.

4   Proposal

We consider an existing AI service system developed for an enterprise function into a different function in a different industry. As described in Sect. 3.1, An enterprise function comprises some business tasks and we introduce an AI service system to (semi-)automate such a business task. An AI application service for the task represented in the OODA model is shown in Fig. 3. The practitioners execute a business task by referring to the outcomes derived from facts by the

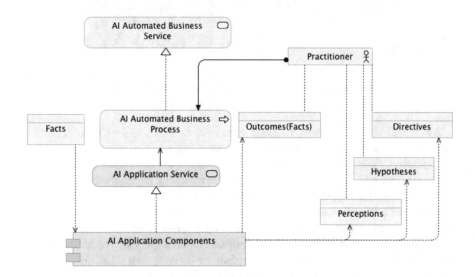

Fig. 3. AI service system used in the OODA process

AI application components. They can also refer to the intermediate outputs (perceptions, hypotheses, or directives) if necessary.

Next, we derive a model for AI application components for implementing an AI application service in Fig. 3. Most of the business functions in companies are assumed to be knowledge-intensive tasks. A model representing the knowledge intensive data-processing system (KIDS) is proposed in [5], which comprises the following four components:

- Transformation component: Extract and convert data required for the business task.
- Assessment component: Structure information according to the objectives of the business task.
- Resolution component: From the structured information, find possible options for the outcomes of the business task.
- Enactment component: Determine the outcomes of the possible options based on the task-specific preconditions.

Based on this KIDS model, AI application components are represented by an ArchiMate visualization, as shown in Fig. 4. Each application component links to the model of a specific function and these models are represented as data objects. The definitions of data objects are shown in Table 2.

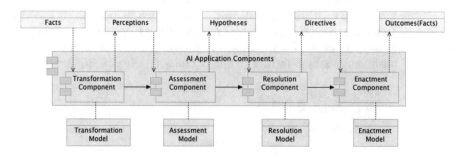

Fig. 4. AI application components and data objects

It is found that an AI service system comprises four application components represented in Fig. 4, which are realized using machine-learning engines, logic programming, or general software products. When we introduce an existing AI service system into our own business task, we expect to minimize the cost of adopting the service system for the new domain.

Now, for the application components (semi-)automating the OODA process, we derive the necessary condition that the component can work for the inputs in the new business task. From the definitions of business objects in the OODA model and data objects in the KIDS model, we identify the conditions required for each OODA process or KIDS component. Table 3 shows the identified conditions.

Table 2. Data objects in the KIDS model

Data Object	Definition
Transformation Model	Model for extracting and converting data required for the task from the original data
Assessment Model	Model for constructing the structured information specific for the task
Resolution Model	Model for identifying the output options from large candidates
Enactment Model	Model for determining the final output

Table 3. Conditions required for the OODA processes and the KIDS components

Business Process	Application Component	Condition
Observation	Transformation	C1: Input to the business task is directly provided into the application component. C2: Quality of input data is same.
Orientation	Assessment	C3: The structured information for the objective of the business task and its schema is same.
Decision	Resolution	C4: All options required to execute the new business task are represented.
Action	Enactment	C5: The final output can be provided to the user through the interface.

In the adoptability assessment, we assess whether the sample input works for these application components and sequentially confirm whether the C1 and C5 conditions are satisfied with the sample input. If one of the conditions is not satisfied, we assume that the AI application service will stop at the corresponding component and provide information representing that the AI application service does not work properly for the input to the user.

5   Experiment

To confirm whether we can assess the adoptability of AI systems by using the proposed method without support of data scientists, we conduct an experiment using an AI speaker. We assume that the project members without deep knowledge on AI can assess the adoptability of AI systems if they can predict the behavior of AI systems in their own business context by themselves.

Based on this assumption, we define the experimental environment using an AI speaker, which is called a smart speaker and a type of wireless speaker with a virtual assistant, providing information based on the voice query from users. In our experiment, we use Google Home mini[2] as the AI speaker. Though an AI speaker is a product for general use, and sometimes, cannot work for a query based on the user's specific needs, users without any experience of AI systems

---

[2] https://store.google.com/product/google_home_mini.

tend to ask questions freely under their own context. In this experiment, we confirm that users can predict whether their queries can work for the AI speaker, by using the proposed method.

As subjects of our experiment, we set two groups of university students (Groups A and B). Each subject group comprises seven students who do not have any experience of AI speakers or other AI-assistant services. We show a list of queries asked to the AI speaker to the subjects in each group, who then assess whether the AI speaker works for these queries. Table 4 shows the example queries in the list.

Table 4. Example queries

Example query
What is the current train operation status ?
What is the trend fashion ?
How many world heritages are there in Brazil ?
Which shops are having a sale ?

Each subject predicts whether the AI speaker can respond to the query appropriately. In this prediction, they assign one of the following response types to each query.

(i) AI speaker can respond to the query appropriately.
(ii) AI speaker responds such as "I cannot understand."
(iii) AI speaker responds such as "I most likely won't be helpful for you" or evades the subject.

The subjects in both groups assess 26 queries in the list (Dataset 1) in their own way and predict the response types. Then, we send these queries to the AI speaker and obtain the actual response types. By comparing the predicted responses with the actual ones, we calculate the prediction correctness score S as follows.

$$S = \frac{\text{\# of queries where the predicted type is same with the actual type}}{\text{\# of queries}} \tag{1}$$

In case (i), we do not care about the correctness of the response from the AI speaker. We calculate the average scores for Groups A and B on Dataset 1 and represent them as S_{A1} and S_{B1}, respectively.

Next, we explain the proposed model of AI service systems, the conditions for the OODA processes, and the KIDS components required in AI service systems to the subjects in Group A. After these explanations, we substantiate the conditions for the AI speaker with them. Table 5 represents the derived conditions. In this situation, all subjects in Groups A and B assess another 26 queries in the list (Dataset 2) and predict the response types again. This means that the subjects in Group A assess the system with the knowledge on the model of AI service systems. By following the same procedure, we calculate the average scores for Groups A and B on Dataset 2 and obtain S_{A2} and S_{B2}. Table 6 shows the evaluation results of the response prediction. In this table, $(\cdot)$ represents the variance of scores in each group for each dataset. This result indicates that the prediction of the response type is notably improved in Group A compared to Group B by introducing the proposed adoptability assessment method.

Table 5. Substantiated conditions for AI Speakers

Business Process	Application Component	Substantiated condition for AI speakers
Observation	Transformation	C1: Input is voice by user. C2: Used in a quiet space like a room in a house.
Orientation	Assessment	C3: User's query consists of the intention and the entity (object of interest).
Decision	Resolution	C4: Information of the options for the query is explicitly described on the Web.
Action	Enactment	C5: The output can be expressed by voice.

Table 6. Substantiated conditions for AI speakers

	Dataset 1	Dataset 2
Group A	$S_{A1}=0.643$ (0.000984)	$S_{A2}=0.721$ (0.00305)
Group B	$S_{B1}=0.623$ (0.0177)	$S_{B2}=0.675$ (0.00236)

6   Discussion

From the experiment result, though it is found that the prediction correctness score is improved in Group A, the score is also improved slightly in Group B. The difference between the scores in Group B comes from the different query sets. To validate this, we conduct two statistical tests (ST_1 and ST_2), where we consider the following null hypotheses:

$$\text{(Statistical test } ST_1) \qquad \text{H}_0 : S_{A1} = S_{B1} \qquad (2)$$

$$\text{(Statistical test } ST_2) \qquad \text{H}_0 : S_{A2} = S_{B2} \qquad (3)$$

In both ST_1 and ST_2, the sample size is the number of members, which is the same in both groups. Therefore, we can apply the t-test under the condition that the two sample sizes are the same and the two distributions have different variance. Under this condition, the t statistic to test whether the means are the same is calculated as 0.0558 in $ST1$. In ST_2, the statistic is also calculated as

1.63. In both tests, these statistics follow the t distribution with $12(= 7 + 7 - 2)$ degrees of freedom. Therefore, the p-values of ST_1 and ST_2 are 0.478 and 0.0640, respectively. At the statistical level of $0.10(10\%)$, H_0 is not rejected in ST_1 and H_0 is rejected in ST_2. This means that we cannot support the fact that the sampled data for S_{A2} and those for S_{B2} come from the distributions with the same mean.

These statistical tests indicate that predicting the response type of the AI speaker is improved by using the proposed adoptability assessment method. However, in our experiment, only 14 university students, as experimental subjects, are considered. We need to define more subjects with wide varieties of backgrounds. Additionally, from this result alone, we cannot argue that the improvement is sufficient and anyone without deep knowledge on the technologies used by AI speakers can understand the capability of AI speakers, such as data scientists. For this purpose, we need to know how much correctly data scientists predict the behavior of the AI speaker, using the dataset in our experiment. This will be considered in a future study.

In our experiment, we use AI speakers having a question-answering function, which is a well-known AI technology. Some AI service systems used for supporting enterprise functions comprise some specific minor AI technologies. For such systems, we need to investigate whether we can apply the proposed method effectively. This investigation is also our future work.

7   Conclusion

In this study, we considered the adoption of AI service systems developed in one business task to a different business task. In some businesses, there is a demand in business divisions to introduce the existing AI service systems into their own business area effectively and easily. In such a situation, we must assess the adoptability of existing AI service systems without deep knowledge, skills, and experience on AI technologies. For this purpose, we proposed a method to assess whether we can introduce an existing AI service system to the target business task. Through the OODA and KIDS modeling, we identified conditions for the adoption of the AI service system. Based on the experiment using an AI speaker, we confirmed that we can predict whether the example inputs work for the AI speaker using the proposed method. From the statistical tests conducted on the experimental results, it was found that we can predict the behavior of the AI speaker using the proposed adoptability assessment method. As a future study, we need to extend the AI service systems used in the experiment to investigate whether our method can effectively work for the adoptability assessment of AI service systems in a real business domain.

Acknowledgements. This work was supported by JSPS Grants-in-Aid for Scientific Research (KAKENHI) Grant Number JP19K20416.

References

1. Amershi, S., Begel, A., Bird, C., Deliner, R., Gall, H., Kamar, E., Nushi, N.N.B., Zimmermann, T.: Software engineering for machine learning: a case study. In: Proceedings of the 41st International Conference on Software Engineering, pp. 291–300 (2019)
2. Baclawski, K., Chan, E.S., Gawlick, D., Liu, Z.H., Ghoneimy, A., Gross, K., Zhang, X.: Framework for ontology-driven decision making. Appl. Ontol. **12**(3–4), 245–273 (2017)
3. Demchenko, Y., de Last, C., Membrey, P.: Defining architecture components of the big data ecosystem. In: Proceedings of the International Conference on Collaboration Technologies and Systems (CTS), pp. 104–112 (2014)
4. Earley, S.: Analytics, machine learning, and the internet of things. IEEE ITPro **17**(1), 10–13 (2015)
5. Gawlick, D., Chan, E.S., Ghoneimy, A., Liu, Z.H.: Mastering situation awareness: the next big challenge? SIGMOD Rec. **44**(3), 19–24 (2015)
6. Heit, J., Liu, J., Shah, M.: An architecture for the deployment of statistical models for the big data era. In: Proceedings of IEEE International Conference on Big Data, pp. 1377–1384 (2016)
7. Hinkelmann, K., Gerber, A., Karagiannis, D., Thoenssen, B., van der Merwe, A., Woitsch, R.: A new paradigm for the continuous alignment of business and IT: combining enterprise architecture modelling and enterprise ontology. Comput. Ind. **79**, 77–86 (2016)
8. Kim, M., Zimmermann, T., DeLine, R., Begel, A.: The emerging role of data scientists on software development teams. In: Proceedings of the 38th International Conference on Software Engineering, pp. 96–107 (2016)
9. Lu, J., Behbood, V., Hao, P., Zuo, H., Xue, S., Zhang, G.: Transfer learning using computational intelligence: a survey. Knowl.-Based Syst. **80**, 14–23 (2015)
10. Saat, J., Franke, U., Lagerström, R., Ekstedt, M.: Enterprise architecture meta models for IT/business alignment situations. In: Proceedings of the 14th IEEE International Enterprise Distributed Object Computing Conference, pp. 14–23 (2010)
11. Sternberg, R.J.: Successful Intelligence: How Practical and Creative Intelligence Determines Success in Life. Simon & Schuster, New York City (1996)
12. Takeuchi, H., Yamamoto, S.: Business AI alignment modeling based on enterprise architecture. In: Proceedings of the 11th KES International Conference on Intelligent Decision Technologies (Springer Smart Innovation, Systems and Technologies vol. 143), pp. 155 – 165 (2019)
13. The Open Group: ArchiMate 3.0.1 – A Pocket Guide. Van Haren Publishing (2013)

Changes in Industry-Academia-Government Collaboration Using the Regional Development Process Model for Information Systems

Takayuki Kunieda[1(✉)], Satoru Yamada[2], Yusuke Kometani[1], Naka Gotoda[1], and Rihito Yaegashi[1]

[1] Faculty of Engineering and Design, Kagawa University, Saiwaicho 1-1, Takamatsu, Kagawa 760-8521, Japan
{kunieda.takayuki,kometani.yusuke,gotoda.naka, yaegashi.rihito}@kagawa-u.ac.jp
[2] RICOH Co., Ltd., Nakamagome 1-3-6, Ota-ku, Tokyo 143-8555, Japan
satoru.yamada@jp.ricoh.com

Abstract. Kagawa University has jointly developed a regional problem-solving information system with Ricoh Co., Ltd. These are Advertising Display Printer System (KadaPos), Travel Diary Generation/Printing System (KaDiary) and Tourist Guidebook Generating/Printing System (KadaPam). These information systems were developed by combining Ricoh's image matching technology and print control technology with the information system development technology of Kagawa University. These systems were developed based on the "Kagawa-type development process model". The model is composed of "extraction of regional problems", "planning solutions for problems", "development of prototype systems to verify solutions" and "demonstration experiments using prototype systems". In this paper, we describe the practice of the "Kagawa-type development process model" in the development processes of KadaPos, KaDiary and KadaPam. The practice of regional problem-solving information system development based on "Kagawa-type development process model" shows that there is a certain effect in the development of regional problem-solving information system and technology verification in enterprises. Furthermore, as a result of advancing three development examples in accordance with the "Kagawa-type development process model," industry-academia-government collaboration changed. This paper describes the effectiveness of the "Kagawa-type development process model" and the resulting changes in industry-academia-government collaboration.

Keywords: Industry-academia-government collaboration · Kagawa-type development process model · Regional problem-solving information system · Open innovation

1 Introduction

So far, many Japanese companies have been mainly using development methods that use only in-house management resources from research called closed innovation to product

© Springer Nature Switzerland AG 2020
M. Virvou et al. (Eds.): JCKBSE 2020, LAIS 19, pp. 96–108, 2020.
https://doi.org/10.1007/978-3-030-53949-8_9

development. "Closed innovation" has been shown to be effective in accumulating R & D budgets/periods in the company and accumulating intellectual property. Globalization is progressing and competition among companies is intensifying. The expansion of the market due to the economic growth of emerging countries has generated diversification of the market needs required for products. In addition to responding to such needs, the product development based on new ideas not found in other companies, New development methods different from the conventional ones are required. "Open innovation" refers to technologies, ideas, services, know-how, data, other companies, universities, local governments, social entrepreneurs, etc. It is an innovation methodology that combines knowledge, etc., and leads to innovative business models, research results, product development, service development, organizational reform, administrative reform, regional revitalization, social innovation, etc., and is proposed by Henry Chesbrough of Harvard Business School [1].

In the information system development and software development of waterfall type development, in order to develop a high quality and high reliability information system and software, the requirements are determined at the early stage of development, and the development can proceed according to the determined requirement specifications. Customer needs are diversified by rapid development of information and communication technology. In order to develop competitive information systems and software, other development methods different from waterfall type development, such as agile type development, have been proposed for Web application and Web system development. Agile development is a development method that tries to minimize risk by adopting development in short time units called iteration. In agile development, in order to develop software in a short period of time, it is common to determine the Application Programming Interface (API) and develop it using the Software Development Kit (SDK), which is a package that bundles necessary programs and documents. In recent years, SDKs that can be used via networks have emerged and become widely used due to the speedup of networks including mobile communications and the improvement in performance of mobile terminals. TAMAGO Labs [2, 3] has developed an SDK that can be used on the network to promote "Open Innovation" with Ricoh's image matching technology and print control technology. Companies are promoting Open Innovation not only for widely using their own technology to solve regional problems, but also for judging the effectiveness and marketability of technology at an early stage.

Kagawa University has jointly developed a regional problem-solving information system with Ricoh Co., Ltd. These are Advertising Display Printer System (KadaPos), Travel Diary Generation/Printing System (KaDiary) and Tourist Guidebook Generating/Printing System (KadaPam). These information systems were developed by combining Ricoh's image matching and printing technologies and Kagawa University's information system development technologies. The development process consists of "extraction of regional problems", "planning of solutions to issues as hypotheses", "development of prototype systems for verifying hypotheses", "demonstration experiments using prototype systems" and "demonstration experiment results", "business judgment based on". This process is based on Kagawa-type development process model. The structure of this paper is shown below. Section 2 describes the Kagawa-type development process model.

Section 3 describes information system examples developed based on the Kagawa-type development process model. Section 4 gives a summary of this paper.

2 Kagawa-Type Development Process Model

This chapter describes the Kagawa-type development process model. The Kagawa-type development process model is a model for developing regional problem-solving information systems that combine lean startup and agile development processes based on the idea of open innovation. This is the first proposal in this paper. Figure 1 shows the process flow of the Kagawa-type development process model. The Kagawa-type development process model consists of 8 stages. In this model, parties such as industry, government, and academia that participate in problem solving are called members.

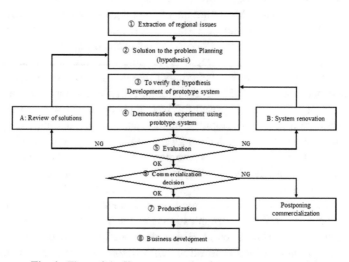

Fig. 1. Flow of the Kagawa-type development process model

(1) In extracting regional problems, interviews and fieldwork with local residents, local governments, and companies are conducted to extract regional problems and share the extracted regional problems with members.

(2) As a hypothesis, in planning solutions to problems, members share what technologies and ideas they have and how they can be used. In addition, the technology that is lacking will be clarified. In order to advance development through open innovation, it is important to recognize the resources of members who work together on issues. Next, on the premise that these resources are available, members will devise solutions. In the development of problem-solving information systems, the creation of solutions is important, but at Kagawa University we use an idea method that uses an analogy based on design thinking to create solutions. However, this idea method is not stipulated in the Kagawa-type development process model.

(3) In the development of the prototype system for verifying the hypothesis, the prototype system is developed using agile development in order to minimize the risk. This prototype system adopts an agile development method that can flexibly respond to specification changes in the middle, not a waterfall type that establishes requirements at the beginning of development, and adopt evolutionary prototyping [4].

(4) In the "demonstration experiment using the prototype system", the demonstration experiment is conducted with the cooperation of local governments, companies and universities.

(5) In the evaluation, if there is an error in the solution itself as a result of the proof experiment, return to (2) "Drafting a solution to the problem as a hypothesis" (A in the figure). If there is a problem with the system, such as a deficient function, functional defect, or failure in the prototype system, return to (3) "Development of a prototype system to verify the hypothesis" (B in the figure).

The cycle from (2) to (5) is implemented as a process that saves speed and waste as much as possible as the practice of MVP (Minimum Viable Product) [5] in the concept of lean startup. In the Kagawa-type development process model, the university is in a leading position in clarifying the roles of all members and managing the overall progress. In the Kagawa-type development process model, the issue is how to disclose the technology that has been made available in-house to the outside. Ricoh Co., Ltd. did not initially have an interface for use outside of its own products. By developing a regional problem-solving information system based on the Kagawa-type development process model, the company was promoted to shift from a closed system to an open system.

Figure 2 shows the configuration of the conventional closed system. Ricoh Co., Ltd. has developed multifunction printers, video conference systems, interactive whiteboards, and so on. Each device has been designed to be a common part within the company, but all functions had an interface for internal use. In addition, the interface for internal use frequently changes in specifications due to functional improvements, and it was not possible to use it without understanding complex specifications and conditions.

Figure 3 is a configuration diagram of an open system in which technology that has been used only for its own products is widely disclosed to many users. In order to publish the technology owned by a company in this way, it is important to determine the API of the technology provided and prepare an SDK that can be used easily and easily by the user. By providing the functions according to the API, the technology can be provided to developers regardless of the specification changes that frequently occur in the company. Furthermore, by simplifying complex specifications and preparing easy-to-understand APIs, developers who develop information systems can easily use them. Ricoh Co., Ltd. provides APIs for the technologies it provides in order to encourage the shift to open systems that make its technology widely available to many users through the development of regional problem information systems based on the Kagawa-type development process model. We have provided an SDK that is easy for users to use.

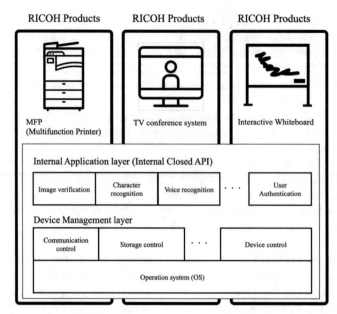

Fig. 2. Configuration of closed system

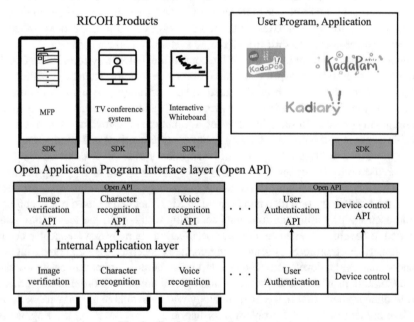

Fig. 3. Configuration of open system

3 Practice of Kagawa-Type Development Process Model

Kagawa University has jointly developed a regional problem-solving information system with Ricoh Co., Ltd. These are Advertising Display Printer System (KadaPos), Travel Diary Generation/Printing System (KaDiary) and Tourist Guidebook Generating/Printing System (KadaPam). These systems were developed using the Kagawa-type development process model. In this chapter we describe the practice of the Kagawa-type development process model.

3.1 Advertisement Display Printer System (KadaPos)

KadaPos [6] is a printing system that provides students with regional information by printing the regional information (Information on local shopping streets, information on events held in the area, information on local professional sports organizations, etc.) according to the student's attributes on the back of the paper used for educational research activities. Students can use KadapPos for free. Figure 4 shows the system outline of KadaPos. KadaPos is a printing system that prints regional information according to student attributes on the back of print paper. KadaPos has developed a system using Ricoh's printing control technology. The printing control technology provided by Ricoh was originally developed as an SDK for internal use, but the functions provided as an SDK for KadaPos development were reviewed (provided by public API). Figure 5 shows the transition of the method for providing the print control API. The initial KadaPos released in January 2017 provided an API developed as an in-house closed system. Therefore, to control the printer, it was necessary to prepare a server and to develop it by understanding more complicated API. An addition, the API specifications provided in response to improvements in printer functions were frequently changed, resulting in development that could only be handled by Ricoh engineers. In the next KadaPos, we placed servers with print control technology on the cloud and focused on the functions used by system developers. As a result, the development efficiency of the system was improved, and the developer was able to develop the system using the print control technology without Ricoh engineers. KadaPos has developed the following process using the Kagawa-type development process model.

(1) There was a problem raised by local governments regarding the revitalization of shopping streets in Takamatsu City as "extraction of regional problems".

(2) Kagawa University proposed a solution that connects students and shopping streets via a printing service in "Proposing solutions (hypotheses) to problems". In response to this proposal, Ricoh Co., Ltd. provided its own printing technology.

(3) "Development of prototype system for testing hypotheses" was started in 2015.

At Kagawa University, KadaPos was in the phase of (5) "Evaluation" as of March 2018, after repeating (4) "demonstration experiment using prototype system" and "system renovation".

(5) As a result of "evaluation", it was found that there was a need for a copy function (a function to copy the manuscript brought in by the user and add advertisement information on the back side). Returning to (4), "Proof experiment using prototype system" has been conducted since January 2019.

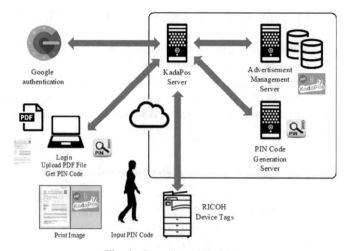

Fig. 4. Overview of KadaPos

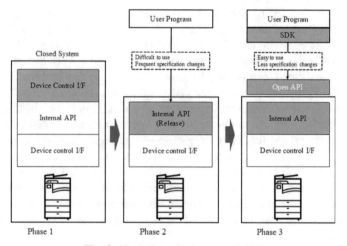

Fig. 5. Evolution of print control API

As of July 2019, it was found that many users were using the copy function by going to the (5) "evaluation" phase, so we are considering business in the (6) "Business Decision" phase.

The qualitative change in industry-academia-government collaboration resulting from the development of KadaPos has changed the way companies provide technology publicly. In addition, companies provide technology using the SDK with a simple and easy-to-understand API, allowing students to develop prototype systems.

3.2 Travel Diary Generation/Printing System (KaDiary)

KaDiary [7] is a system that generates a travel diary on an electronic medium from photographs taken by tourists during sightseeing and prints them using a printer. KaDiary extracted requests from interviews that cannot collect the tourist behavior of tourists in Shodoshima-Town. Next, it was developed as a mechanism that allows tourists to remember sightseeing from the extracted issues and collects tourists' sightseeing behavior. It is a system that collects tourist behavior information from EXIF information (GPS location information and time information) of photos taken by tourists, generates a tourist diary, prints it, and provides it to tourists. In printing, as with KadaPos, a server with print control technology was placed on the cloud, and development was performed using an API (printer control P3 in Fig. 5) focused on the functions used by system developers. Figure 6 is an overview of KaDiary's system. KaDiary consists of a tourist information transmission application, a tourist information registration application, and Ricoh's print control technology. Like KadaPos, KaDiary does not require a Ricoh technician. The system could be developed using print control technology.

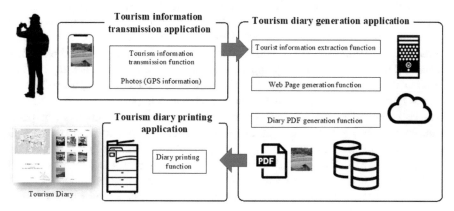

Fig. 6. Overview of KaDiary

KaDiary has developed the following process using the Kagawa-type development process model.

(1) Kagawa University interviewed Shodoshima-Town for "extraction of regional problems".

(2) "Planning solutions (hypotheses) for issues" was conducted by Kagawa University. In response to a proposal from Kagawa University, Ricoh Co., Ltd. provided an improved printing technology using KadaPos.

(3) "Development of prototype system for verifying hypotheses" was conducted, and (4) "Verification experiment using prototype system" was conducted in Shodoshima island in November 2016.

KaDiary finished the phase of (5) "Evaluation" and considered business in the phase of (6) "Judgment of business", but gave up business. However, the opinion that the tourist behavior of tourists can be extracted has been received, and the members are considering whether to conduct the (6) "Business Decision" phase again.

As a qualitative change in industry-academia-government collaboration, local governments will solve local problems by conducting demonstration tests in the field provided by local governments using a prototype system developed by students through the development of KaDiary. Increasingly involved in activities.

3.3 Tourist Guidebook Generating/Printing System (KadaPam)

KadaPam [8] is a system that generates a personalized tourist guidebook that records the memories of the trip by replacing the tourist guidebook photos with photos taken by the tourists themselves. The generated guidebook can be printed using a printer connected to the cloud. Figure 7 shows an overview of the KadaPam system. KadaPam was developed by applying gamification [9] in order to actually invite tourists who could not be realized by the conventional tourist guidebooks to the tourist spots listed in the tourist guidebooks. Gamification refers to activities that apply game design and game principles to other than games. A progress bar that shows the progress of a task is a mechanism that promotes task progress by presenting the target element to the user and visualizing the progress of the target element, and is one example of an application of gamification. For stamp rally and photo rally in tourism, stamps are installed at the sightseeing spots (presentation of target elements) and collected around the sightseeing spots where the stamps are set (visualization of progress). Stamp rally and photo rally are efforts to promote tourism by applying gamification, and it has been reported that there is a certain effect on the increase in the number of visited tourist spots and the stay time of tourist spots. KadaPam uses tourist guidebooks to present the target elements to tourists and replaces the original photos posted in the tourist guidebooks with photos taken by the tourists themselves at the same location and composition (progress By using image matching technology, we have realized a mechanism that facilitates task execution (visit to tourist destinations) by visualizing. In other words, KadaPam actually invites tourists who could not be realized with conventional tourist guidebooks to the tourist spots listed in the tourist guidebooks and looks at the tourist guidebooks to check whether tourists actually visited the tourist spots It was developed with the aim of doing. KadaPam was developed using not only Ricoh's printing control technology but also the image matching technology RICOH TAMAGO Snapi SDK [10] (Snapi SDK). The Snapi SDK compares the arrangement of many feature points (image feature patterns) of the object and background existing in the registered key image with the query image input by the image matching technology developed based on the character recognition technology. Thus, it is a technique to determine whether the same image feature pattern exists in multiple places. KadaPam uses this image matching technology to determine whether a photograph in the tourist guide book is a photograph taken with the same composition as a photograph taken by a tourist. When it is determined that they are the same, the tourist guide book is replaced with a photograph taken by the tourist. The Snapi SDK technology was originally developed for use in products that use character matching developed by Ricoh Co., Ltd. (closed system). TAMAGO Labs. Has decided to publish the API as an SDK and deploy the image matching engine on the cloud so that this technology can be used in a wide variety of fields. The development of KadaPam was advanced using this image matching technology and print control technology. KadaPam has developed the following process using the Kagawa-type development process model.

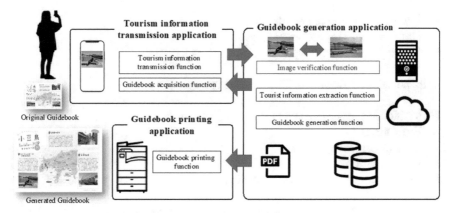

Fig. 7. Overview of KadaPam

(1) As for "extraction of regional problems", we noticed that there are many cases where paper tourist guidebooks are thrown away after use at tourist sites in Kagawa Prefecture.

(2) Kagawa University proposed to replace the photos in the tourist guide book with their own photos as "planning solutions (hypotheses) for the issues". To achieve this, Ricoh Co., Ltd. provided its own image matching technology in addition to printing technology.

In 2017, (3) "Development of prototype system to verify hypotheses" was conducted.

KadaPam conducted (4) "demonstration experiment using prototype system" in Shodoshima in November 2017 and (5) performed "evaluation".

As a result, it was judged that there was no problem with the solution to the problem and the developed information system, and it proceeded to the phase of "6".

(6) Even in the "business decision" phase, it is judged that there is a prospect of business in terms of market and funds. Currently (7) We are implementing initiatives aimed at business in the "Production" phase.

In the development of KadaPam, as a result of the proof experiment, not only the local government that provided the place for problem solving, but also other local governments offered to use it. As a qualitative change in industry-academia-government collaboration, RICOH has begun the challenge to commercialize KadaPam as a result of several municipalities offering to use KadaPam.

4   Summary

Kagawa University developed Advertising Display Printer System (KadaPos), Travel Diary Generation/Printing System (KaDiary) and Tourist Guidebook Generating/Printing System (KadaPam) through industry-government-academia collaboration with Ricoh Co. Ltd., Takamatsu City, and Shodoshima Town. These information systems were developed by applying the Kagawa-type development process model consisting of the following processes.

(1) Extraction of regional problems
(2) Draft solutions (hypotheses) for issues
(3) Development of prototype system for testing hypotheses

   In the fields of Takamatsu City and Shodoshima Town

(4) Verification experiment using prototype system
(5) Evaluation
(6) Business decision
(7) Productization
(8) Market introduction/business development

In the case of developing information services to solve local issues, the development using the Kagawa-type development process model was effective in the following ways. The reasons for this are that, when promoting activities in industry-academia-government collaboration, it becomes clear at what stage the activity is currently at, and what needs to be done next can be shared by all sides of the industry-academia-government.

When applying the Kagawa-type development process model to solve regional issues through industry-academia-government collaboration, the spiral-up described below in industry-academia-government collaboration began to turn around.

Initially, the relationship between industry, academia, and government was such that local governments outsourced business, such as financial support for companies and universities, companies outsourced research to universities, and companies outsourced business based on research results. However, by applying the "Kagawa-type development process model" to solve local issues and turning the process, companies began to disclose information so that their technologies could be used at universities. As a result, universities and corporations have endeavored to tackle local issues endogenously, and companies have changed to make their technologies more open and accessible (open innovation). This trend has brought the advantage that students can develop a prototype system to develop to verify the effectiveness of the regional problem solution. At universities, students will be able to develop a prototype system, which will speed up and lower costs for conducting demonstration experiments, and will enable a faster turnaround (spiral-up) to solve and verify the issues identified in the demonstration experiments. Even from the point of view of local governments, such movements have become more realistic as companies and universities have realized the realization of initiatives for solving local issues. As a result, instead of providing competitive funding as in the past, it has shifted to participating in specific initiatives on regional issues that are governed by local governments. In addition, the initiative of local governments has brought concrete solutions to local issues at companies and universities, and this attitude has encouraged companies to look for business feasibility. If business is successful, the solution to the regional problem will be implemented as a business rooted in the region, and if there is a gap between the solution and the regional problem or if further improvement is required, the local university will promote it A virtuous cycle has come to be created, such as working on the creation of new solutions.

Figure 8 conceptually illustrates a virtuous cycle of industry-academia-government collaboration. In this way, the "Kagawa-type development process model" has helped to create a good relationship between industry, academia and government and a spiral-up of activities for solving regional issues.

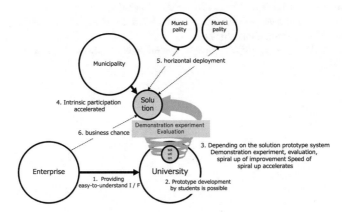

Fig. 8. Industry-academia-government collaboration model

This paper described the Kagawa-type development process model and the practice of the Kagawa-type development process model in the development of KadaPos, KaDiary, and KadPam. The practice of regional problem type information system development at Kagawa University based on the Kagawa-type development process model shows that the Kagawa-type development process model has a certain effect on the development of regional problem solution type information system. Currently, a new information system that solves regional problems is under development.

References

1. Ministry of Education, Culture, Sports, Science and Technology: 2017 White Paper on Science and Technology "What is Open Innovation?" (2017). http://www.mext.go.jp/b_menu/hakusho/html/hpaa201701/detail/1388436.htm. Accessed 16 Oct 2018
2. RICOH TAMAGO Labs. https://www.ricoh.co.jp/software/tamago/. Accessed 06 Dec 2018
3. Kunieda, T.: Challenge to "making new products" in companies Introduction of development example of SDK (RICOH TAMAGO Snapi), a content registration/search service using "key images". In: Proceedings of the 2017 IEICE Society Conference, BI-3-3. IEICE, Tokyo (2017)
4. Guida, G., Lamperti, G., Zanella, M.: Software Prototyping in Data and Knowledge Engineering. Springer, Dordrecht (2013). 408p
5. Ries, E.: Lean Startup. Nikkei BP, Tokyo (2012)
6. Takada, R., Gotoda, N., Kii, M., Kunieda, T., Yamada, S., Sano, H., Takeshita, Y., Yaegashi, R.: Development of a advertisement printer system (KadaPos). IPSJ Digit. Pract. **8**(4), 325–333 (2017)
7. Kumano, K., Miyagawa, R., Kunieda, T., Yamada, S., Gotoda, N., Kii, M., Yaegashi, R.: Development of travel diary generating/printing system (KaDiary). IPSJ Digit. Pract. **8**(4), 316–324 (2017)

8. Kunieda, T., Izumi, R., Miyagawa, R., Ikeda, T., Kanaya, M., Yamada, S., Gotoda, N., Kometani, Y., Yaegashi, R.: Development of Tourist guidebook generating/printing system (KadaPam) that records travel memories and analysis of travel behavior using travel guidebook in Shodo-shima. IPSJ Digit. Pract. **10**(4), 829–849 (2019)
9. Cho, Y.-K.: Actionable Gamification: Beyond Points, Badges, and Leaderboards. Packt Publishing Ltd., Birmingham (2019)
10. RICOH TAMAGO Snapi SDK. https://www.ricoh.co.jp/software/tamago/snapi/. Accessed 07 Feb 2018

Identifications of Webpage Layout Patterns Based on Lines of Vision

Junko Shirogane[1(✉)], Atsuko Eguchi[2], Hajime Iwata[3], and Yoshiaki Fukazawa[2]

[1] Tokyo Woman's Christian University, Tokyo, Japan
junko@lab.twcu.ac.jp
[2] Waseda University, Tokyo, Japan
bleu3oiseau0w0@akane.waseda.jp, fukazawa@waseda.jp
[3] Kanagawa Institute of Technology, Kanagawa, Japan
hajimei@nw.kanagawa-it.ac.jp

Abstract. Webpage layouts strongly affect usability. Although usability has various elements, this study focuses on readability. To realize a good readability, some layout patterns are well known by webpage contents. Because users do not always see the intended layout patterns by web designers, readability may not be realized. Thus, we propose a method to identify the implemented layout patterns for target webpages using the users' lines of vision. In our method, blocks that consist of the layout patterns and the order of the blocks, which users see, are specified for a webpage. Users' lines of vision are analyzed when viewing the webpages, and the implemented layout patterns are identified. Concretely, whether an implemented layout pattern agrees with the intended one is analyzed. If the recorded coordinates of the lines of vision are included in the specified blocks and the blocks are viewed in the specified order, then the implemented and intended layouts of the webpage agree.

Keywords: Webpage layout · Line of vision · Eye tracker

1 Introduction

Webpages consist of various blocks such as headers, footers, contents, and side menus. Headers are placed at the top, while footers are on the bottom. Contents and side menus are placed between headers and footers. Typically, they are arranged side-by-side (two-column layout) or only contents (one-column layout). In addition, colors are used. Base colors, main colors, and accent colors are specified by the rules of the webpages.

The layout and colors affect the usability of webpages. We focus on users' lines of vision for webpage usability. Various studies have proposed lines of vision in webpages such as call to action (CTA) [1], Z-shaped pattern [2], and F-shaped pattern [3]. CTA means the lines of vision are directed to an action point (mainly click buttons that webpage designers want users to click). In the Z-shaped pattern, users initially see the top block of a webpage from left to right. Next, they

© Springer Nature Switzerland AG 2020
M. Virvou et al. (Eds.): JCKBSE 2020, LAIS 19, pp. 109–121, 2020.
https://doi.org/10.1007/978-3-030-53949-8_10

see the webpage from the top right to the bottom left followed by the bottom block from left to right. In the F-shaped pattern, users see the top block of a webpage from left to right. Next, they see the next block of the top block from left to right followed by the left block of the webpage from top to bottom. CTA is a webpage layout where designers navigate users to action points, while Z- and F-shaped patterns are layouts for readability (layout patterns). Our study considers the readability.

In arranging webpages, web designers assume the users' lines of vision and often decide the arrangement of above blocks based on patterns (intended patterns). However, even if web designers assume patterns such as F- or Z-shaped ones, the users' lines of vision do not always follow the intended patterns. This may cause issues with the readability of the webpage.

Thus, we propose a method to identify the layout patterns that users see (implemented patterns) by analyzing users' lines of vision. In the proposed method, blocks on the target webpages corresponding to the intended patterns are specified first. Then the coordinates of the blocks are acquired. The users' lines of vision are assumed to move in the blocks. Next, users see the target webpages, and their lines of vision are recorded. Afterwards the lines of vision included in the specified blocks and the order they see the blocks are analyzed. Finally, the layout patterns based on users' lines of vision are identified. This allows webpage designers to evaluate whether the intended and implemented patterns correspond to each other.

2   Related Works

Various studies have employed data analysis on lines of vision in webpages and the usages of lines of vision. For lines of vision data analysis, Falkowska et al. verified the validity of a usability test using an electroencephalogram (EEG) [4]. Their study analyzed the correlation between the characteristics of users' lines of vision and emotional reactions when users saw web applications. Users conducted four tasks in the target web application, and their lines of vision characteristics and emotional reactions were recorded. To evaluate users' emotional reactions, the analysis was performed by four web elements (area of interests, AOI). They used SimplyTick (an E-commerce plugin for online stores) [5] as the target web application, Tobii X2-60 [6] as the eye tracker, and Emotiv Epoc EEG [7] to obtain emotional reactions. Consequently, they revealed part of the correlations between the lines of vision characteristics and emotional reactions such as those between short-term excitement and the average time of access duration and between long-term excitement and average time of lines of vision. Menges et al. proposed a method to represent webpage screenshots to analyze lines of vision data [8]. The target webpages had fixed and scrollable elements. Their study aimed to resolve accuracy and scalability issues when representing lines of vision data on webpages. They initially identified and cropped fixed elements on webpages. Next, a screenshot of all scrollable elements was taken. Finally, screenshots of the fixed elements and scrollable elements were combined.

Although these studies described the relations between lines of vision data and webpages, they did not focus on the order of blocks in webpages that users see.

For the usage of lines of vision data, Clark et al. evaluated whether mouse tracking could be substituted for eye tracking because eye-tracking hardware and software are expensive [9]. They aimed to find correlations between users' eye lines of vision and mouse cursor positions when viewing the target webpages. Thus, the participants saw pages of news, courses, and publications in the target websites. They analyzed the recorded lines of vision data and mouse movements using the Florida Institute of Technology Human-Centered Design Institute as the target website, Eye Trible Eye Tracker as the eye tracker [10], and EyeProof [11] as the analysis software. Consequently, the number of mouse accesses and number of lines of vision were strongly correlated between all target webpages and the participants.

Eraslan et al. evaluated the effects of user number to analyze users' eye movement sequences (scanpaths) [12]. Participants browsed the target webpages for 30 seconds and conducted search tasks. Their lines of vision data were analyzed by the authors' analysis algorithm for scanpath and improved Vision-based Page Segmentation (VIPS) algorithm [13]. For searching and browsing tasks, 27 and 34 of the 65 users had 75% similarity, respectively. In these studies, although users' lines of vision were traced, they did not consider whether the intended and implemented layouts were the same.

3   Layout Patterns

Layout patterns are content arrangement based on users' lines of vision for webpages. Considering the lines of vision on webpages, Z-shaped pattern [2] and F-shaped pattern [3] are known to exhibit a good readability. These patterns are effective for webpages with a lot of text and regular arrangement of contents, such as arrangement of goods in a matrix on a shopping web page. In this section, we describe these two patterns. However, our method is not limited to these patterns. In addition, when webpages have attractive factors such as colors and large-sized elements, these factors do not always contribute to readability [14,15]. Our targets are webpages that follow the layout patterns and do not have attractive factors.

3.1   Z-Shaped Pattern

In the Z-shaped pattern, users first see the top of the contents from left to right horizontally. Next, they move their lines of vision from the top right to the bottom left. Then they see the bottom of the contents from left to right horizontally. The users' lines of vision trace the shape of the letter "Z" on the target webpages. Figure 1[1] shows an example of tracing a webpage by a user's line of vision.

---

[1] https://www.amazon.com/, (Accessed 19 March, 2020).

In many cases, users see the first visited webpages by the Z-shaped pattern and grasp the abstract structures of the target webpages. Examples of webpages suitable for the Z-shaped pattern are advertising and poster pages as well as the top pages of electronic commerce sites.

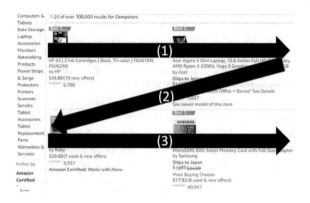

Fig. 1. Example of tracing a webpage by the Z-shaped pattern

3.2 F-Shaped Pattern

In the F-shaped pattern, users see the top of the contents in the target webpage from left to right horizontally. Then they see the next contents of the top contents from left to right horizontally. Finally, they see the left contents from top to bottom vertically. The users' lines of vision trace the shape of the letter "F" on the target webpages. Figure 2[2] shows an example of tracing a webpage by a user's line of vision.

In many cases, users see webpages by the F-shaped pattern when they understand the details of the webpage contents after grasping the abstract structures. Examples of webpages suitable to the F-shaped pattern are blogs and the introduction of products on electronic commerce sites.

4 Identifications of Webpage Layout Patterns

Our method acquires users' lines of vision using eye trackers and identifies the implemented patterns that users see. Web designers compare the implemented patterns with the intended patterns. Then they evaluate whether their implementations are appropriate. Figures 3 and 4 overview our method and details of our system, respectively.

---

[2] https://www.nngroup.com/articles/f-shaped-pattern-reading-web-content-discovered/, (Accessed 19 March, 2020).

4.1   Lines of Visions

Users' lines of vision data are acquired using eye trackers. Eye trackers can trace movements of lines of vision and record coordinates of places on displays where users gaze and the times by specific frequency. Our method uses the coordinates and times to identify the implemented patterns of webpages.

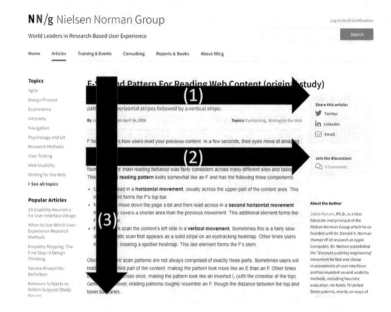

Fig. 2. Example of tracing a webpage by the F-shaped pattern

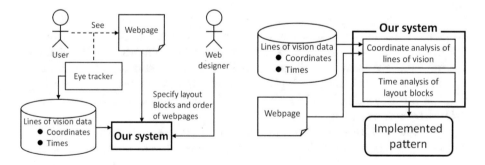

Fig. 3. Method overview   **Fig. 4.** System detail

4.2   Webpage Layout Patterns

Our method analyzes whether users see the intended pattern for the target webpages. First, to analyze the intended pattern of the target webpage, blocks, which represent layout patterns (layout blocks), must be specified. Blocks correspond to areas represented by tags on the webpage. Web designers specify these blocks by HTML (Hyper Text Markup Language) tags and the order that users see. Figure 5[3] shows an example of the layout block specification. Then our system analyzes the coordinates of the layout blocks on the webpage.

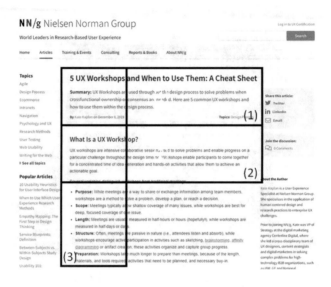

Fig. 5. Example of a layout block specification

Second, our system extracts the coordinates of lines of vision recorded by the eye tracker included in the layout blocks. If the rate of coordinates outside the layout blocks is less than the threshold, our system evaluates that the shape of the implemented pattern is the same as that of the intended pattern. Because the layout block size and user types drastically differ from webpage to webpage and the threshold is affected by the block size and user types, the threshold is defined by web designers.

The layout pattern is defined by the order that users see each layout block. Third, for a certain layout block, the average time for the recorded coordinates is calculated. After calculating the average times for all layout blocks of the layout

---

[3] https://www.nngroup.com/articles/5-ux-workshops/, (Accessed 24 March, 2020).

pattern, they are used to analyze the order of layout blocks that users actually see. If the order of the layout blocks that users see based on the average times of the layout blocks is the same as that of the intended pattern, our system evaluates that users see the webpage using the intended pattern.

5   Implementation

To implement our method, we used Tobii Pro nano [16] as an eye tracker and Google Chrome DevTools [17]. Tobii Pro nano, which is an eye tracker of the screen base, records the coordinates of users' lines of vision at a 60-Hz frequency. Using Tobii Pro Lab Full Edition [18], which is a software package of Tobii Pro nano, heat maps for the gazing time of the lines of vision and gaze maps for movement of lines of vision can be recorded. Google Chrome DevTools is a developer tool of Google Chrome that analyzes the source programs of webpages and acquires the coordinates of tag elements on webpages. Although these programs are used in this study, our method can be implemented using other tools. The requirements for eye trackers are recording coordinates and times of users' lines of vision, while the requirements for web browsers are analyzing source programs and acquiring coordinates of tag elements on webpages.

5.1   Coordinates of Tag Elements

After the web designer specifies the layout blocks that represent the intended pattern on the target webpage, tags corresponding to the specified layout block are extracted using Google Chrome DevTools. For example, the corresponding tags for (1) layout block in Fig. 5 is "h1". Based on the extracted tags, the coordinates of the layout blocks are identified.

5.2   Layout Pattern

While viewing the target webpage, users' lines of vision are recorded. Then the lines of vision are analyzed, and the implemented patterns are identified based on their lines of vision. The basic identification strategies are described in Sect. 4.2. In this section, we describe the identification strategies of individual layout patterns. Currently, our targets are Z- and F-shaped patterns. However, if web designers specify the layout blocks and orders for other layout patterns that users' lines of vision move on a webpage, our method is applicable to other layout patterns.

Z-Shaped Pattern. When the recorded coordinates satisfy the following three criteria, our system evaluates that the users see the Z-shaped pattern. Our criteria are explained using the arrow in Fig. 1:

Z-a Users initially see (1) layout block in Fig. 1 followed by (3) layout block.

Z-b When users' lines of vision move from (1) to (3), the movement is from the right of (1) to the left of (3).

Z-c The rate of the number of recorded coordinates by users' lines of vision outside (1) and (3) in Fig. 1 is less than the specified threshold in Sect. 4.2.

Criterion Z-a is satisfied if the average time of the recorded coordinates in (1) occurs before that of (3) layout block. Criterion Z-b is satisfied if the recorded coordinates outside (1) and (3) are between (1) and (3) and move from up to down and from right to left. Criterion Z-c is satisfied if the calculated rate of the recorded coordinates outside (1) and (3) layout blocks in Fig. 1 is less than the threshold.

F-Shaped Pattern. When the recorded coordinates satisfy the following three criteria, our system evaluates that the users see the F-shaped pattern. Our criteria are explained using the layout blocks in Fig. 5:

F-a Users initially see layout block (1) followed by layout block (2).

F-b After users see layout block (2), they see layout block (3).

F-c The rate of the number of recorded coordinates by users' lines of vision outside (1), (2), and (3) in Fig. 5 is less than the specified threshold in Sect. 4.2.

For criteria F-a and F-b, the average times of the recorded coordinates in (1), (2), and (3) layout blocks are calculated. Criteria F-a and F-b are satisfied if the average times follow the order of (1), (2), and (3). For criterion F-c, the rate of the recorded coordinates outside (1), (2), and (3) layout blocks in Fig. 5 is calculated. Criterion F-c is satisfied if the rate is less than the threshold.

6   Evaluation

To evaluate our method, we conducted experiments where participants viewed 20 webpages. Then the implemented patterns were identified based on their lines of vision by our method and the results were compared to the intended patterns. The participants consisted of 8 males and 7 females ranging in age from 21 to 24. The experiment involved webpages matching the concepts in Sects. 3.1 and 3.2. Half of the webpages were Z-shaped patterns (Webpage Z-

1 to Z-10)[4, 5, 6, 7, 8, 9, 10, 11, 12, 13], and the other half were F-shaped patterns
(Webpage F-1 to F-10)[14, 15, 16, 17, 18, 19, 20, 21, 22, 23].

6.1 Experimental Design

The experiment involved three steps:

1. Because the movements of users' eyes vary by individual, the eye tracker was adjusted for the user's eyes by the calibration function.
2. The eye tracker recorded the lines of vision as users viewed a webpage. This step was repeated for all target webpages.
3. Our system identified the layout patterns by users' lines of vision.

As described in Sect. 4.2, when the rate of the number of the recorded coordinates of a user's lines of vision outside of the layout blocks is below a certain threshold, our system assesses that a user sees the intended pattern. Although the web designer can set the threshold, we specified the threshold in this experiment. To specify the threshold, we viewed a target webpage along with the layout pattern in an arbitrary manner carefully at once, and our lines of vision were recorded. That is, we viewed a target webpage of Z- and F-shaped patterns by Z- and F-shaped patterns, respectively. The rates of the coordinates of the lines of vision outside the layout blocks was calculated. The rates were specified as the baseline values (BVs), and we identified the layout patterns of the target webpages using three types of thresholds: BV-5%, BV, and BV+5%.

---

[4] https://t.pia.jp/ (Accessed 3 February, 2020).

[5] https://www.amazon.co.jp/ (Accessed 3 February, 2020).

[6] https://kate.ocnk.net/ (Accessed 3 February, 2020).

[7] http://mishima-home.com/home.com/ (Accessed 3 February, 2020).

[8] https://gori.me/ (Accessed 3 February, 2020).

[9] https://www.isa-school.net/ (Accessed 3 February, 2020).

[10] https://covers.holiday/ (Accessed 3 February, 2020).

[11] https://www.orangepage.net/ (Accessed 3 February, 2020).

[12] https://www.three-count.jp/ (Accessed 3 February, 2020).

[13] http://happy-room.net/ (Accessed 3 February, 2020).

[14] https://www.nngroup.com/articles/5-ux-workshops/ (Accessed 3 February, 2020).

[15] https://web-doctor.jp/column/ (Accessed 3 February, 2020).

[16] http://toudai5000.net/?p=66 (Accessed 3 February, 2020).

[17] https://www.kids.pref.ibaraki.jp/kids/home/xs=_WP.zS8YFLg/,   (Accessed   3 February, 2020).

[18] http://komotaro.blog.jp/archives/39293330.html#more   (Accessed   3   February, 2020).

[19] https://pharm-kusuri.com/copy/z.html (Accessed 3 February, 2020).

[20] https://covers.holiday/products/detail/174 (Accessed 3 February, 2020).

[21] https://gucciz03.com/toreaku-335 (Accessed 3 February, 2020).

[22] https://www.excite.co.jp/news/article/FrenchBulldogLife_ID36139/   (Accessed   3 February, 2020).

[23] https://www.makeshop.jp/main/know-how/attract/attract-point.html, (Accessed 3 February, 2020).

6.2   Experiment Results

We counted the number of participants that viewed the target webpages with the Z-shaped pattern for Z-1 to Z-10 and the F-shaped pattern for F-1 to F-10. Table 1 shows the results for the Z-shaped pattern. For F-1 to F-10, because many of the participants did not view the F-shaped pattern, we analyzed the participants' lines of vision and evaluated whether the participants satisfied two of the three criteria for identifying F-shaped pattern in Sect. 5.2. Table 2 shows the results of the F-shaped pattern where the participants satisfied all of the criteria to identify the F-shaped pattern, while Table 3 shows the results of the F-shaped pattern where participants satisfied two of the three criteria. These tables show BV and the rates of participants who saw the webpages along with the layout patterns.

7   Discussion

For the threshold, when the values are large, our system evaluates that more participants viewed the webpages with the intended patterns. As described in Sect. 4.2, the threshold is defined by web designers. Although web designers can try various thresholds using our method, it is difficult to define the appropriate threshold. Thus, Appropriate threshold definition must be supported by our method.

Over half of the participants did not see the Z-shaped pattern for Z-2, Z-5, Z-8, Z-9, and Z-10 even when the threshold was set to BV+5%. To confirm the reason, we analyzed the gaze maps of these webpages, which represent the order and the time length of points where participants gaze. Although the participants saw these webpages along with Z-shaped pattern, the gaze points were

Table 1. Result of seeing webpages with the Z-shaped pattern

		Threshold		
	BV	BV − 5%	BV	BV + 5%
Z-1	74.81	26.67	26.67	53.33
Z-2	47.80	0.00	6.67	20.00
Z-3	36.13	26.67	66.67	93.33
Z-4	69.38	26.67	53.33	60.00
Z-5	55.45	26.67	40.00	46.67
Z-6	43.37	26.67	53.33	66.67
Z-7	66.67	20.00	40.00	66.67
Z-8	28.97	13.33	13.33	40.00
Z-9	66.81	0.00	0.00	0.00
Z-10	75.14	0.00	0.00	6.67

Table 2. Result of seeing webpages with F-shaped pattern (All criteria were satisfied)

	BV	Threshold		
		BV − 5%	BV	BV + 5%
F-1	65.96	6.67	6.67	6.67
F-2	96.16	13.33	26.67	26.67
F-3	37.10	0.00	0.00	0.00
F-4	73.91	13.33	13.33	13.33
F-5	34.47	0.00	6.67	6.67
F-6	50.55	0.00	0.00	0.00
F-7	52.81	6.67	13.33	26.67
F-8	34.94	13.33	13.33	13.33
F-9	51.44	6.67	6.67	6.67
F-10	79.84	0.00	0.00	13.33

Table 3. Result of seeing webpages with F-shaped pattern (two criteria were satisfied)

	BV	Threshold		
		BV − 5%	BV	BV + 5%
F-1	65.96	33.33	33.33	40.00
F-2	96.16	60.00	86.67	86.67
F-3	37.10	53.33	53.33	53.33
F-4	73.91	33.33	53.33	80.00
F-5	34.47	53.33	73.33	73.33
F-6	50.55	13.33	53.33	60.00
F-7	52.81	40.00	86.67	100.00
F-8	34.94	13.33	20.00	20.00
F-9	51.44	20.00	20.00	20.00
F-10	79.84	0.00	6.67	20.00

not included in the specified layout blocks. To identify these webpages as the Z-shaped pattern, strategies to identify layout patterns without specifying layout blocks must be proposed. Even though participants saw these webpages with Z-shaped pattern, web designers did not assume the Z-shaped pattern. Thus, it is possible that the readability of these webpages should be improved. We must discuss which options should be adopted.

Over half of the participants identified the F-shaped pattern for F-12, F-13, F-14, F-15, F-16, and F-17 when the threshold is BV+5% using two criteria (Table 3). However, less than half viewed the F-shaped pattern using all three criteria (Table 2). This difference is attributed to the time comparison. In the F-shaped pattern, layout blocks (1) and (2) in Fig. 5 overlap with (3). That is, many coordinates of lines of vision are used in plural layout blocks when calculating average times of layout blocks. To resolve this problem, the calculation of the average time or the specification of layout blocks must be improved.

Even when using only two criteria, more than 50% of participants did not see the F-shaped pattern for F-1, F-8, F-9, and F-10. Hence, we analyzed gaze maps and heat maps of these webpages. The heat maps represent the staying time length of lines of vision. Participants saw the F-shaped pattern, but the overlapping areas strongly affected these webpages.

According to these results, there are some problems of the layout block specifications and average time calculations. However, Z-1, Z-3, Z-4, Z-6, F-2, F-3, F-4, F-5, F-6, and F-7 were correctly evaluated by more than 50% of participants using a threshold of BV+5%. Considering that BV is the value of seeing the target webpages in an arbitrary manner, our method identifies the layout patterns of webpages appropriately.

8 Conclusion

To realize readability, web designers arrange web contents based on various layout patterns and assume that users view webpages using the intended patterns. However, the intended and implemented patterns often differ. Here, we proposed a method to identify the implemented patterns that users saw. Our method recorded and analyzed users' lines of vision to determine the implemented pattern. Then the implemented pattern was compared to the intended one. In our method, web designers specified the intended layout blocks and viewing order. Our system analyzed whether the coordinates of the lines of vision were included in the layout blocks and whether the average times of the coordinates in the individual layout blocks followed the intended order. If the rates of the coordinates outside the layout blocks were below a certain threshold and the average times did not follow the intended order, our system evaluated that users view webpages following the intended pattern.

As an evaluation, we selected 20 webpages. Then 15 participants viewed these webpages and whether the participants saw these webpages using the intended patterns was analyzed. Although there were some issues with the layout block specifications and average time calculations, our method tended to appropriately identify the layout patterns.

Future works include:

- Increasing the number of layout patterns that our system can identify;
- Improving the average time calculation for overlapping layout blocks;
- Identifying layout patterns statistically.

References

1. Liu, D., Cao, Y., Lau, R.W.H., Chan, A.B.: ButtonTips: design web buttons with suggestions. In: Proceedings of 2019 IEEE International Conference on Multimedia and Expo (ICME) (2019)
2. Lidwell, W., Holden, K., Butler, J.: Universal Principles of Design, Revised and Updated: 125 Ways to Enhance Usability, Influence Perception, Increase Appeal, Make Better Design Decisions, and Teach through Design Second Edition, Rockport Publishers (2010)
3. Nielsen, J.: F Shaped Pattern For Reading Web Content (original study) (2006). https://www.nngroup.com/articles/fshapedpatternreadingwebcontentdiscovered/discovered/. Accessed 3 Feb 2020
4. Falkowska, J., Sobecki, J., Pietrzak, M.: Eye tracking usability testing enhanced with EEG analysis. In: Proceedings of International Conference of Design, User Experience, and Usability (DUXU 2016) (2016)
5. SimplyTick, Lantan Software. https://apps.shopify.com/simplytick. Accessed 5 April 2020
6. User's Manual Tobii X2-60 Eye Tracker. https://www.tobiipro.com/siteassets/tobii-pro/user-manuals/tobii-pro-x2-60-eye-tracker-user-manual.pdf/?v=1.0.3. Accessed 5 April 2020
7. EMOTIV EPOC+, https://www.emotiv.com/epoc/. Accessed 5 April 2020

8. Menges, R., Tamimi, H., Kumar, C., Walber, T., Schaefer, C., Staab, S.: Enhanced representation of web pages for usability analysis with eye tracking. In: Proceedings of the 2018 ACM Symposium on Eye Tracking Research & Applications (ETRA 2018) (2018)

9. Clark, J.W., Stephane, A.L.: Affordable eye tracking for informed web design. In: Proceedings of International Conference of Design, User Experience, and Usability (DUXU 2018) (2018)

10. The Eye Trible. https://theeyetribe.com/theeyetribe.com/about/index.html. Accessed 5 April 2020

11. The Eye Tribe Launches Cloud-based Eye Tracking Analytics Platform. https://www.prnewswire.com/news-releases/the-eye-tribe-launches-cloud-based-eye-tracking-analytics-platform-249273631.html. Accessed 5 April 2020

12. Eraslan, S., Yesilada, Y., Harper, S.: Eye tracking scan path analysis on web pages: how many users? In: Proceedings of the Ninth Biennial ACM Symposium on Eye Tracking Research & Applications (ETRA 2016) (2016)

13. Akpinar, M.E., Yesilada, Y.: Vision based page segmentation algorithm: extended and perceived success. In: Proceedings of International Conference on Web Engineering (ICWE 2013) (2013)

14. Tidwell, J.: Designing Interfaces: Patterns for Effective Interaction Design, 1st Edn. O'Reilly Media, Sebastopol (2005)

15. Lidwell, W., Holden, K., Butler, J.: Universal principles of design: 100 ways to enhance usability, influence perception, increase appeal, make better design decisions, and teach through design by Jill Butler, Rockport Publishers (2007)

16. Tobii Pro Nano. https://www.tobiipro.com/product-listing/nano/. Accessed 5 April 2020

17. Chrome DevTools. https://developers.google.com/web/tools/chrome-devtools. Accessed 5 April 2020

18. Tobii Pro Lab. https://www.tobiipro.com/product-listing/tobii-pro-lab/. Accessed 5 April 2020

STAMP S&S: Layered Modeling
for the Complexed System in the Society
of AI/IoT

Tomoko Kaneko$^{(\boxtimes)}$ ⓘ and Nobukazu Yoshioka ⓘ

QAML Project, National Institute of Informatics, Tokyo, Japan
{t-kaneko,nobukazu}@nii.ac.jp

Abstract. Systems, including AI/IoT, have complex relationships. It is necessary to analyze risks from various perspectives to build a system that can be used safely and securely throughout society, including people and organizations. Object modeling is desirable for risk analysis from multiple viewpoints.

An accident model based on system theory called STAMP and its hazard analysis method STPA has attracted attention recently. The basis of this theory is the Control Structure diagram (CS) that captures the entire system. The authors use CS as a structural diagram that captures the requirements of the whole system, including humans and society, and clarifies the relationship by the software life-cycle process standard and the system-life cycle process standard. Therefore, it is proposed to explain the specifications hierarchically for each software, system, service, and stakeholder, and to standardize it for the social layer. In order to model a complex system hierarchically, we propose to model the control structure diagram of STAMP into five layers according to the life cycle of software and system requirements. In addition, we present a case study of safety and security analysis based on the above-mentioned model, considering the case of level 3 autonomous driving.

Keywords: STAMP · Socio-technical system · ISO21207 · ISO15288 · Layered modeling · Risk analysis · Safety · Security

1 Introduction

The IoT (Internet of Things), which connects things and people through computer systems, has had a considerable effect on society. Not only are things connected, but systems, people who use them, and organizations have a complex relationship [1]. Although the evolution of AI (Artificial Intelligence), especially deep learning, is being rapidly commercialized, it is challenging to ensure the safety and reliability of machine learning systems, including mission-critical systems such as automatic driving.

To ensure safety and reliability in the age of AI/IoT, it is necessary first to understand the entire target system, and the effects of its components on each other, model them clearly and deal with risks. However, it is difficult to model a complex system as a whole that its implementation has not been established. Currently, the life cycle model standards

© Springer Nature Switzerland AG 2020
M. Virvou et al. (Eds.): JCKBSE 2020, LAIS 19, pp. 122–131, 2020.
https://doi.org/10.1007/978-3-030-53949-8_11

for systems and software specify the requirements for "what" to be implemented during the planning, development, operation, and maintenance processes. However, the problem is that "it does not specify how to do it." Engineers working on systems that require complex and diverse considerations, including human and social, such as autonomous driving and smart cities, want to know how to do it. As a first step in solving this problem, the authors aim to "establish a way to model complex systems for risk analysis." Traditionally, the use of a single device, a component of a complex system, has been analyzed as much as possible. However, this method has its limitations for the risk analysis of a complex system as a whole. And it is desirable that the modeling is in accordance with development standards such as the software life cycle model.

Therefore, this paper presents four issues for modeling complex systems. As a countermeasure, we propose to model the control structure diagram of the accident model STAMP [2] based on system theory into five layers according to the life cycle of software and system requirements. (This is called STAMP S & S). Furthermore, it is argued that applying STAMP's hazard analysis method called STPA [3] to risk analysis can be a specific means to extract risks according to each layer, through examples studied in level 3 autonomous driving. In this paper, the related knowledge is introduced inSect. 2, the STAMP S & S is proposed in Sect. 3 to solve the problem of complex system analysis and its solution from the beginning, and Sect. 4 considers the case of the level 3 autonomous driving.

2 Related Work

2.1 STAMP and Its Related Methods

Modern embedded systems are becoming gradually larger and more complexed due to the interaction among connected elements in addition to the advanced functionality of each element. Traditional safety analysis techniques (such as Fault Tree Analysis [4], FMEA [5], and HAZOP [6]) are based on accident chain event models. However, since this model seeks causal relationships between individual events, it cannot capture a complex system as a whole. Therefore, to ensure the safety of these complex systems, Leveson proposed the System Theory Accident Model and Process (STAMP) and its analysis methods, like System Theory Process Analysis (STPA). The mechanism of STAMP is explained by focusing on the element (component) and the interaction (Control action) in the Fig. 1. Many of the system accidents are not only caused by the failure of the components, but also by the interaction of the control elements (control element and the controlled element) for safety in the system. As a process, STAMP use specifications, safety guide design, design principles, system engineering, risk management, management principles, and regulation of organizational design. However, STAMP is originally an accident model for safety and has not been established as a method for analyzing the impact of computer systems on society.

2.2 The Standards of Software Life Cycle Processes and System Life Cycle Processes

Software life cycle processes (ISO/IEC/IEEE 12207) [7] provides processes that can be employed for defining, controlling, and improving software life cycle processes within

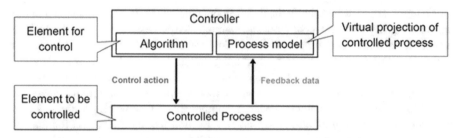

Fig. 1. Control and controlled component in STAMP [5]

an organization or a project. The processes, activities, and tasks of this document can also be applied during the acquisition of a system that contains software, either alone or in conjunction with ISO/IEC/IEEE 15288 [8]. The ISO/IEC 15288 is a systems engineering standard covering processes and lifecycle stages. In the context of this document and ISO/IEC/IEEE 15288 [8], there is a continuum of human-made systems from those that use little or no software to those in which software is the primary interest. It is rare to encounter a complex system without software, and all software systems require physical system components (hardware) to operate, either as part of the software system-of-interest or as an enabling system or infrastructure. Thus, the choice of whether to apply this document for the software life cycle processes or ISO/IEC/IEEE 15288:2015 [8] depends on the system-of-interest. Processes in both documents have the same process purpose and process outcomes but differ in activities and tasks to perform software engineering or systems engineering, respectively. Also, while these standards provide information on what processes can be used by systems and software, they do not indicate how to take concrete measures.

2.3 Social Technology System and Software Engineering

The system, which includes nontechnical elements such as people, processes, and regulations, as well as technical components such as computers, software, and other equipment, is called Sociotechnical System [9].

A socio-technical system includes hardware, software, people, and organizations. Socio-technical systems are so complex that it is impossible to understand them as a whole.

Therefore, you have to view them as layers. The socio-technical systems stack is shown in Fig. 2. Software systems are not isolated systems but are part of more extensive systems that have a human, social, or organizational purpose. Therefore software engineering is not an isolated activity but is an intrinsic part of systems engineering. Also, as shown in Fig. 2, software engineering includes business process, application system, communication and data management, and operating system layers, and system engineering covers organization and equipment in addition to them. But society alone is not included in either. In addition, although a complex system requires modeling based on such a hierarchy, it does not show how to analyze it.

Fig. 2. The socio-technical systems stack [9]

3   STAMP S & S (Modeling Based on System Theory)

3.1   Issues and Countermeasures for Complex System Analysis

As mentioned before, modeling a complex system as a whole is difficult, and implementation has not been established. By considering the context of standards, socio-technical systems, and software engineering, we think complex system analysis has the following four issues. In this paper, we propose STAMP S & S as hierarchical modeling to solve these four issues.

Issue 1: Overview of Complexed System
Issue 1 is that "complex systems cannot be captured as a whole."
    As a countermeasure against this, the CS diagram itself is based on the system theory and gives a bird' s-eye view of the entire system. The reason that this method is valid is that STAMP is a methodology that captures a complicated system as a whole as an accident model based on system theory. Also, a complex system was modeled, as shown in Fig. 2.

Issue 2: The Need for Hierarchical Modeling
Issue 2 is that "Computer systems affect not only systems but also various objects, but the objects are various and complicated, such as people and organizations, in addition to the included software. Socio-technical systems are so complex that it is impossible to understand them as a whole. Rather, you have to view them as layers" [9].
    As a countermeasure, it is necessary to model a complicated system hierarchically. As shown in Fig. 2, it was possible to model with layers.

Issue 3: Life Cycle Process Standards and Layered Modeling
Issue 3 is that the analysis that captures the entire system including people and organizations is performed using the STAMP model, but it is not defined how the model is divided into layers for detailed analysis.
    As a countermeasure for Issue 3, STAMP S & S divides the software layer, system layer, service layer, and stakeholder layer according to the system life cycle process standard [7] and software process standard [6]. ISO/IEC/IEEE 12207 already presents a

development process that is divided into software layers, system layers, service layers, and stakeholder layers. This is why you chose this layer. Layers shown in Fig. 2 can also be divided into five hierarchies as follows.

- Organization: Stakeholder Layer
- Business processes: Service Layer
- Application system: Software Layer
- Communications and data management: System and Software Layer
- Operating system: Software Layer
- Equipment: System Layer

Figure 3 shows the relationship between the V-shaped model and stamp S&S and STAMP/STPA. The V-shaped model is divided into four layers from the left end of the V-model: business operations, software, and systems. In the STAMP S & S hierarchy, the business is called the stakeholder layer, and the operations is called the service layer. Systems and software layers are used as they are. The system life cycle process standard [7] is linked with the software life cycle process standard [6], as shown in Sect. 2.

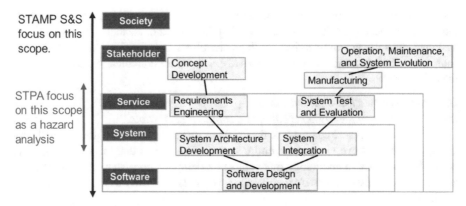

Fig. 3. Relationship between V-shaped model and STAMP S&S and STAMP/STPA

This V-shaped model indicates that, as the development progresses, the corresponding processes usually progress in the order of the layers. However, STAMP S & S's risk analysis is not a waterfall type of risk analysis like this. It performs incremental risk analysis to stay resilient to changing requirements. Also, from the concept stage of development, we intend to conduct a risk analysis that considers the elements of each of these layers. Furthermore, it is intended to carry out a detailed analysis for each layer while considering the interaction. This allows many people to perform various analysis based on one structure.

Issue 4: Social and Technical Systems
Issue 4 is that a method for analyzing the influence of a computer system on society has not been established. Although society is affected by computer systems, it is not possible to identify the responsible party. On the other hand, software, systems, services, and

stakeholders have their responsibility. So standards such as software life cycle processes are subject to integrated analysis. The solution of Issue 4 is to create a method that can analyze, including the social layer, which is not included in software engineering and system engineering [9].

3.2 Significance of Layered Modeling of Complex Systems

The authors extend the use of conventional STAMP not only as an accident model but also as a model that can analyze the impact of complex systems on society. The STAMP model has been shown to adapt to social technology systems, but the structure is not specified in five layers, such as STAMP S&S. In addition, the STAMP model has been mainly subject to analysis of the system layer, service layer, and stakeholder layer presented by STAMP S&S. However, the STAMP model does not make the software layer and the social layer a detailed analysis object.

Although the STAMP model has many advantages, such as the ability to analyze the interaction of multiple components such as people and organizations, we are not only interested in accident models, but also for modeling complex system requirements and risk analysis. The work is proposed as STAMP S & S. S & S is the abbreviation of System, Software, Service, Stakeholder, Society, Specification, Standard, Scenario, Safety and Security. This paper discusses the five layers of S: systems, software, services, stakeholders, and society. The remaining five S evaluate the impact of various events on a scenario basis for safety and security analysis, and show that they will be made into specifications and standards. The relationship between ten S is scheduled to be described separately.

One of the reasons for adding these abbreviations is not only the various devices and systems but also the extended use of STAMP in this article is to model the interaction at the software, system, service, stakeholder and society layers. Its purpose is to build a framework for hierarchical modeling and analysis of complex systems in an AI/IoT society. Although STAMP S & S was presented as a framework for integrating safety and security [10], this paper provides layered modeling as a basis for analyzing safety and security requirements. This modeling includes all the measures to solve the four issues presented in the previous section.

3.3 STAMP S & S and Five Layers

The specific contents of each layer are shown in Table 1. The social layer refers to the laws and regulations of society that govern the operation of the system. It also includes human social life (rules, standards, customs) and its external environment (natural environment such as weather). The stakeholder layer includes the organizational business processes, which make use of the software system, higher-level strategic processes as well as business rules, policies. It is a unit in which a business or organization has a responsibility as a stakeholder. The service layer includes actions performed by people, services, and services provided by people and organizations. The system layer includes a computer system, hardware, communication equipment, semiconductor chip. The software layer includes Programs (application software, OS, and other software), cyber information, data, and AI.

Table 1. Specific contents of each layer

Layer	Contents
Society	Human social life (rules, standards, customs) and its external environment (natural environment such as weather)
Stakeholder	A unit where a business or organization
Service	Actions performed by people, and services provided by people and organizations
System	A computer system, hardware, communication equipment, semiconductor chip
Software	Programs (application software, OS, and other software), cyber information, data, and AI

4   Considerations in Level 3 Autonomous Driving

Autonomous driving is a technology that has a significant impact on society, and Table 2 shows the level of achievement. The control action in the case of the automatic driving level 3 is shown in Fig. 4. In this figure, five layers of STAMP S & S are shown on the left side bylines, and the components of the layers are shown on the right side. In this case, CS shows the interaction between the system layer braking and the control of both the service layer driver and the software layer automatic driving AI. In this way, the hierarchical modeling of STAMP S & S itself is divided into layers and presenting their mutual relations. Also, there are several control actions in this CS diagram, and among them, using the deceleration command to the brake system by the CA5 artificial intelligence module, it is explained that the five layers of STAMP S & S influence each other.

For CA5, unsafe control actions were extracted using the four guide words shown in Table 3. Among them, UCA5-D means "The deceleration command ends before sufficient deceleration is performed, and a collision cannot be maintained with the appropriate distance to the external environment." It is extracted with the guide words, "Stop too soon/Applying too long," which is a hazard. For this hazard, the factors (HCF) were analyzed, as shown in Table 4.

Furthermore, we consider the causal factor that "the external environment could not be recognized due to bad weather.", which is one of the HCFs of (UCA5-D). Bad weather is the external environment and corresponds to the society layer. Artificial intelligence is the software layer. So this factor is affecting the software layer from the society layer. It also shows that measures that span the software and social layers are needed. In SOTIF, the necessity of analysis based on specific conditions (Society) such as climate is mentioned. Therefore, it is conceivable to conduct a detailed analysis while considering these conditions.

As another specific example, "the automatic driving cancellation instruction is falsified, and the automatic driving is terminated during deceleration" can be presented. This is the case when software information is tampered with by a security attack. This is a threat that could be extracted by performing the threat analysis using STRIDE [10, 11]

Table 2. Autonomous driving levels 0 to 5.

Level 0	The driver (human) controls it all: steering, brakes, throttle, power
Level 1	This driver-assistance level means that most functions are still controlled by the driver, but a specific function (like steering or accelerating) can be done automatically by the car
Level 2	At least one driver assistance system of "both steering and acceleration/ deceleration using information about the driving environment" is automated, like cruise control and lane-centering
Level 3	Drivers are still necessary for level 3 cars but are able to completely shift "safety-critical functions" to the vehicle, under certain traffic or environmental conditions
Level 4	This is what is meant by "fully autonomous." Level 4 vehicles are "designed to perform all safety-critical driving functions and monitor roadway conditions for an entire trip
Level 5	This refers to a fully-autonomous system that expects the vehicle's performance to equal that of a human driver in every driving scenario

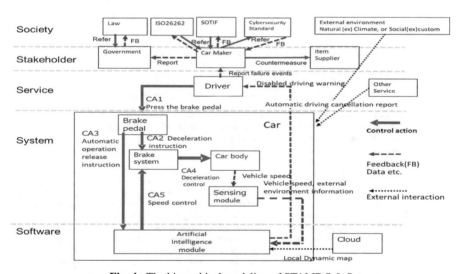

Fig. 4. The hierarchical modeling of STAMP S & S

proposed by STAMP S & S [12]. STRIDE is a threat model in the Microsoft security development lifecycle, as shown in Table 5. STRIDE is based on a reference architecture for determining the overall image of a system. Enumerating threats as a threat analysis diagram is used to verify mitigation and measure mitigations. The purpose of threat modeling, for which it would be bee possible to use a layer, was to understand how an attacker could penetrate the system. This threat affects systems and drivers. In other words, there is an effect on the system layer and the service layer, and detailed analysis

Table 3. Extracting unsafe control actions in CA5

No	CA	Not providing	Providing causes hazard	Too early/Too late	Stop too soon/Applying too long
5	Deceleration command to brake system by artificial intelligence module	(UCA5-N) If artificial intelligence does not issue a deceleration command during automatic driving, it will collide with the external environment [SC1]	(UCA5-P) Unnecessarily strong deceleration command is issued, and rear-end collision occurs [SC3]	(UCA5-T) If the deceleration command is delayed, a collision cannot be maintained with the appropriate distance to the external environment ahead [SC1]	(UCA5-D) The deceleration command ends before sufficient deceleration is performed, and a collision cannot be maintained with the appropriate distance to the external environment [SC1] Continues to issue deceleration command after required deceleration is completed, making acceleration difficult

Table 4. Hazard causal factor analysis for UCA5-D

UCAS	Missing or wrong control input or external information	Inconsistent, incomplete, and inaccurate process models	Inadequate or missing feedback, delayed feedback	Conflicting control actions	Tampering	Denial of service
(UCA5-D)	Unintentional automatic operation release instruction is input from the driver during deceleration	A situation that does not exist in the artificial intelligence learning data has occurred	Speed measured slower than it actually is • The external environment could not be recognized due to bad weather	When the driver senses danger and performs a sudden braking operation, the judgment of artificial intelligence takes precedence	The automatic operation release instruction has been tampered with, and the automatic operation ends during deceleration	The local dynamic map is attacked by a DOS attack, and map reference becomes impossible

and countermeasures are required on both sides. Further, it is up to the stakeholders to consider what means and at what cost should be taken to implement countermeasures to provide such a threat. In addition, the need for countermeasures should be reflected in security standards, which are rules at the social level.

In this way, the safety and security risks analyzed by STAMP S & S are explained in detail at the five layers of the proposal, and a comprehensive measure is needed. By conducting a detailed overview analysis from the concept stage at STAMP S & S, it is possible to consider multilayered measures at an early stage.

Table 5. STRIDE threat model

Threat	Description	Desired property
Spoofing	Involves illegally accessing and then using another user's authentication information	Authenticity
Tampering	Involves the malicious modification of data	Integrity
Repudiation	Associated with users who deny performing an action without other parties having any way to prove otherwise	Non-repudiation
Information disclosure	Involves the exposure of information to individuals who are not supposed to have access to it	Confidentiality
Denial of Service	Denial of service (DoS) attacks deny service to valid users	Availability
Elevation of Privilege	An unprivileged user gains privileged access and thereby has sufficient access to compromise or destroy the entire system	Authorization

5   Conclusion

By modeling and layering software, systems, services, stakeholders, and society, it is proposed that safety analysis of complex systems is possible, and showed the differences and rationale for each of the five layers. Using the case of Autonomous Driving Level 3, we showed where each element of safety and security risks is located in five layers. I want to state that more specific analysis by stratification can be an issue in the future.

Reference

1. Information-technology Promotion Agency (IPA): IoT Safety/Security Development Guidelines
2. Leveson, G.N.: Engineering a Safer World. MIT Press, Cambridge (2012)
3. STPA handbook. https://psas.scripts.mit.edu/home/
4. IEC 61025:2006 Fault Tree Analysis (FTA). https://webstore.iec.ch/publication/4311
5. United States Military Procedure: Procedure for performing a failure mode effect and criticality analysis, MIL-P-1629, 9 November 1949
6. IEC 61882:2001 Hazard and operability studies (HAZOP studies) - application guide. https://www.iec.ch
7. ISO/IEC/IEEE 12207. https://www.iso.org/standard/63712.html
8. ISO/IEC/IEEE15288:2015. https://www.iso.org/standard/63711.html
9. Sommerville, I.: Software Engineering, 10th edn. Pearson Education Limited, London (2016)
10. Lipner, S., Howard, M.: The trustworthy computing security development lifecycle. https://msdn.microsoft.com/en-us/library/ms995349.aspx
11. Shostack, A.: Threat Modeling: Designing for Security. Wiley, New York (2014)
12. Kaneko, T., Takahashi, Y., Okubo, T., Sasaki, R.: Threat analysis using STRIDE with STAMP/STPA. In: The International Workshop on Evidence-based Security and Privacy in the Wild (APSEC 2018 Workshop) (2018)

Parked Car Detection Method Based on Home Wireless LAN -Using Household Equipment to Detect a Parked Car-

Motoko Suzuki$^{(\boxtimes)}$, Yuki Katsumata, and Akira Yamada

NTT DOCOMO, INC., 3-6 Hikari-no-oka, Yokosuka-shi, Kanagawa 239-8536, Japan
{motoko.suzuki.dp,yuuki.katsumata.hs,yamadaakira}@nttdocomo.com

Abstract. Consumer-to-consumer (C2C) car park sharing services are now widespread. In these services, space owners rent their unused land, e.g., an unused parking space next to the owner's home, to a driver. Different from coin parking services, drivers using the park sharing services can make reservations through web sites or mobile-phone applications. There are drivers who abuse this system by, for example, overstaying their reservation. The high cost of parking space sensors that monitor the existence of a parked car makes it cost prohibitive for individuals to install these types of equipment to deter the abuse of parking spaces. To address this problem, we propose a parked car detection technique employing wireless LAN. The proposed method utilizes a household wireless LAN access point and a household laptop computer, and we monitor the received signal strength indicator (RSSI) of beacon frames. This unique method reduces the detection cost and will aid in proliferating car park sharing services. In this paper, we detect the existence of a car in a parking space using the mean and variance of the RSSI. The generated datasets are used for Support Vector Machine (SVM), Decision Tree, and the K-nearest neighbor algorithm (KNN) binary classification. We previously showed that a parked car can be detected when the locations of the equipment are the same. In this study, we show the effect of changing the equipment location in the house on the detection score. These results show that the best recall score is 0.88 and that changing the equipment location does not affect the recall score. The results of this study are expected to be used in a multimodal system.

Keywords: Parking car detection · C2C service · Smart city

1 Introduction

Recently, sharing economy services (hereinafter sharing services) that share idle assets such as space, skills, and goods in a consumer-to-consumer (C2C) manner using the Internet have become popular. The market size of the sharing service market in Japan was 1.9 trillion yen in FY2018, and it is expected to expand to 11 trillion yen by FY2030 [1]. Sharing services match a provider who owns an idle asset with a lessee who wants to use it, and share the idle asset [2]. According to the Ministry of Internal Affairs and Communications of Japan, the percentage of people that have experience using sharing

© Springer Nature Switzerland AG 2020
M. Virvou et al. (Eds.): JCKBSE 2020, LAIS 19, pp. 132–140, 2020.
https://doi.org/10.1007/978-3-030-53949-8_12

services in Japan is 9.1% for car park sharing, 4.9% for ride sharing, 4.9% for house and room sharing, 2.8% for labor sharing such as personal chores, and 1.6% for goods sharing. The car park sharing service has the most users [3]. The use of car park sharing in other countries is higher than in Japan, for example, 16.7% in the United States, 12.3% in Germany, and 14.0% in the UK [3]. The number of users is expected to increase in the future.

In conventional coin parking systems, the user participates in a pay-as-you-go system in which the user pays for the usage time after using the parking space. Therefore, it is necessary to detect the parked car in order to measure the usage time. The two types of detection methods that are widely used are described below. The first is the locking plate method [4]. In this method, a magnetic sensor detects the parked car and a metal plate installed on the ground rises at an angle to bar the car from exiting the parking space essentially locking the car in place. At the end of use, the user inputs the parking space number into the automatic checkout machine and pays the fee. The corresponding metal plate lowers unlocking the car and the user can then drive out of the parking space. The second method is the gate method [4]. The user receives a ticket at the gate of the parking lot entrance. The gate opens and the user proceeds to find a parking space. At the end of use, the user inserts the ticket into the automatic checkout machine and pays the fee. The gate opens and the user can exit the parking lot.

In C2C car park sharing services, the parking space provider rents idle space, so the lessee can use it at a lower cost than for coin parking services. C2C sharing of parking spaces does not employ sensors to detect parked cars or an automatic checkout machine, so users must reserve a parking space and complete payment beforehand. However, there is an issue in which some lessees cannot park even if they reserved a parking space due to other lessees exceeding their reserved time. A dedicated car detection sensor is expensive, and it is not usually installed at home where many of these car parking services are offered. The problem of lessees exceeding their reserved time has hindered the proliferation of car park sharing services.

To address this issue, we investigate a method for detecting a parked car at zero cost using equipment already prevalent in the home [5]. In this study, we focus on detecting a parked car in a parking space that is next to the house using the wireless LAN access point (AP) and a notebook computer already set in the house. We consider a method that detects a parked car using only the Received Signal Strength Indicator (RSSI) of a beacon signal that a laptop computer periodically receives from a wireless LAN AP.

2 Related Research

Security cameras can detect parked cars, but they are not suitable as a method to monitor C2C sharing services because i) the operation cost is high and ii) they give the lessee a sense of third party surveillance imparting a negative impression [6, 7]. In this study, we focus on the sensing method using a radio wave from existing wireless equipment.

Okukawa *et al.* [8] showed that it is possible to detect the state (stationary or moving) of a human from the change in the received power of a beacon signal. In the experiment, a wireless LAN AP and a receiver are set in opposite corners of a room, and human motion is detected in the room. This method is an effective means to detect easily the state (stationary or moving) of a human. However, it is necessary to establish a receiver that comprises dedicated hardware, and it is cost prohibitive for use in sharing services. In addition, this system uses indoor receivers and an AP to detect a human indoors so it is not clear whether it can be applied to outdoor situations such as that for a parked car. In our concept it is necessary for the indoor equipment to detect cars parked outdoors.

Nishi *et al.* [9] proposed a human detection system using the UHF band TV receiving waves from outdoors. However, the radio frequency band is low and the wavelength is approximately 0.4 m to 0.6 m, so the receiver antennas must be large and use dedicated receiving hardware.

Wang *et al.* [10] presented the WiHear system. It detects mouth movements from analyzed fine-grained radio reflections of Wi-Fi signals using Channel State Information (CSI). The CSI is used to listen to people talking, but it requires dedicated equipment.

IEEE802.11 SENS-SG [11] discusses standardization of sensing technology using wireless LAN equipment. Since standardization is based on CSI, we expect that hardware changes will be required.

In order to address these issues in the related research, we aim to establish an outdoor parked car detection technology using wireless LAN communication signals from equipment established indoors.

3 Proposed Method

We propose a method to detect the presence or absence of a parked car using fluctuations in the RSSI value from a beacon signal. The beacon signal is transmitted from a wireless LAN AP that is located indoors, reflected by a parked car outdoors, and received by a laptop computer indoors. This method actualizes sensing without hardware modification because it does not use CSI.

In order to confirm that the proposed method can be used when the layout in the house is changed, we also investigate the effect of changing the location of the equipment.

In the following sections, we describe the detection accuracy of the proposed method in a real environment.

3.1 Experiment Configuration

We receive beacon signals in a two-story house environment with an adjacent parking space as shown in Fig. 1. We consider two experimental conditions. Experiment condition 1 is where a laptop computer is set in room B on the first floor, and a wireless LAN AP is set in room D on the second floor. In experiment condition 2, the locations of the equipment are exchanged as shown in Fig. 2 and Table 1 gives the experimental conditions. The beacon signal is captured on a laptop computer using Wireshark [12]. In the experiment a vehicle enters and exits the parking space 10 times.

Fig. 1. House and parked car used in experiment.

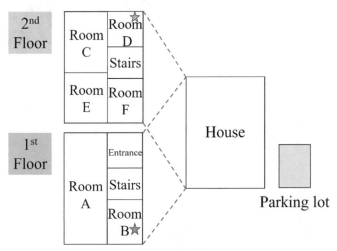

Fig. 2. House layout.

Table 1. Experimental conditions

Equipment	Wireless LAN AP	BUFFALO Air Station
	Laptop computer	Apple MacBook Pro
	Frequency band	2.4 GHz
Size and distance	House	8.3 m × 6.5 m
	Parking space	6 m × 2.5 m
	Parking space and house	0.9 m
Location of equipment	Condition 1	Room B: Laptop computer Room D: AP
	Condition 2	Room B: AP Room D: Laptop computer

3.2 Detection Method

In this section, we consider the detection method. Figure 3 show the RSSI instantaneous values in the environment described in Sect. 3.1. We compare the method using RSSI to that using moving variance and moving average [4]. The results show that the method using the moving variance and moving average exhibits better detection performance.

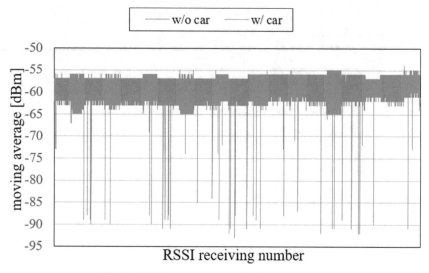

Fig. 3. RSSI instantaneous value for condition 1.

Therefore, in this study we use the moving average and moving variance of the RSSI. Since the RSSI can be acquired in time series, past data are also considered. We use three typical methods of supervised machine learning: Support Vector Machine (SVM), Decision Tree, and K-nearest neighbor (KNN).

We exclude the 1st to 29th data sets. The n-29th to n-st RSSI moving averages and moving variances are calculated. Figures 4 and 5 show the moving average and moving variance in the 8477 data sets for condition 1.

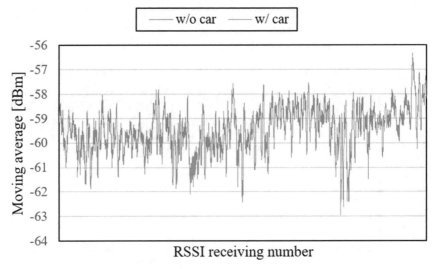

Fig. 4. RSSI moving average for condition 1.

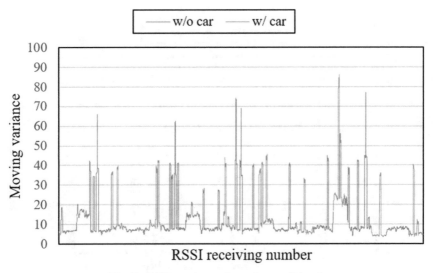

Fig. 5. RSSI moving variance for condition 1.

In this study, we investigate the effect of changing the equipment location in the house on the detection score. Figures 6 and 7 show the moving average and moving variance in the 8497 data sets for condition 2 where the equipment location is changed. The data are divided into training data and verification data in the ratio of 2:8. The training data are put into a classifier and a model is generated. The presence of a parked car is estimated by inserting verification data into the model. The target for the recall score is 0.8 or higher.

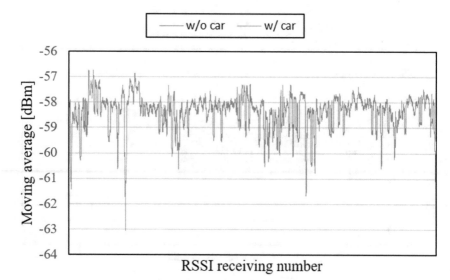

Fig. 6. RSSI moving average for condition 2.

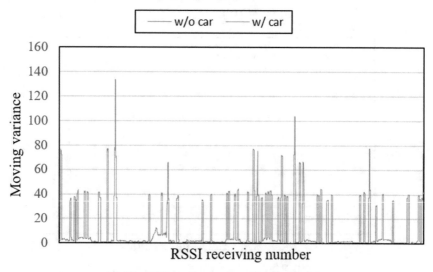

Fig. 7. RSSI moving variance for condition 2.

3.3 Detection Score Comparison

Learned with the test data from each model, the detection score is shown using the evaluation data. The results are shown as Confusion Matrix classes: True Positive (TP), True Negative (TN), False Positive (FP), and False Negative (FN).

$$\text{Accuracy Score} = \frac{\text{TP} + \text{TN}}{\text{TP} + \text{FP} + \text{FN} + \text{TN}} \quad (1)$$

$$\text{Precision Score} = \frac{TP}{TP + FP} \tag{2}$$

$$\text{Recall Score} = \frac{TP}{TP + FN} \tag{3}$$

Figure 8 shows the detection score, precision score, and recall score. Equations (1) to (3) show each performance measurement.

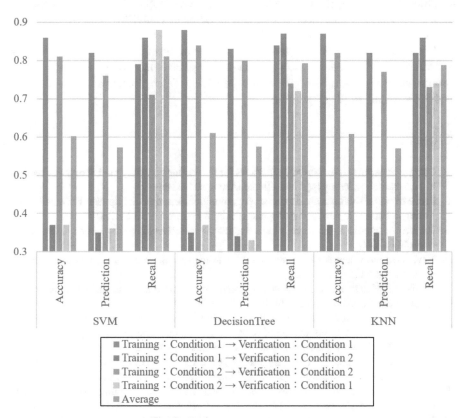

Fig. 8. Performance measurement.

When the learning and evaluation data are obtained under the same conditions (Training: Condition 1 → Verification: Condition 1 and Training: Condition 2 → Verification: Condition 2), better accuracy and prediction scores are obtained than when the learning and evaluation data are obtained under different conditions (Training: Condition 1 → Verification: Condition 2 and Training: Condition 2 → Verification: Condition 1). However, the recall score is high when the training and evaluation data are different. We consider that the characteristics when the vehicle is present are the same.

When compared based on classifier, the best average recall score is 0.81 for SVM. KNN is a classification method based on majority voting of similar data. For that reason it is difficult to use RSSIs that include noise values. The decision tree algorithm may yield poor scores for the test data because the tree becomes excessively deep during the learning process. It is necessary to investigate the depth of the tree in the future.

4 Conclusion

In this study, we detected a parked car in a parking space adjacent to a house to aid in further proliferating the park sharing services. The proposed method analyzed the RSSI of the beacon signal received by a laptop computer established in the house. Using the moving average and moving variance of the RSSI, we showed that the presence or absence of a car can be detected with a maximum recall of 88% when the position of the device is changed. By using the parked car sensing technology and the location information of the reservation person, it is possible to detect illegal parking, which will lead to prevention of illegal parking.

References

1. Sharing Economy Association Japan: News for the sharing economy market research release. https://sharing-economy.jp/ja/news/0409/. Accessed 3 Mar 2020
2. Prime Minister's Office: Sharing economy promotion department. https://www.kantei.go.jp/jp/singi/it2/senmon_bunka/shiearingu/sokushin.html. Accessed 3 Mar 2020
3. Ministry of Internal Affairs and Communications: Research study on realization of integration using ICT (2018)
4. Navi Park: Procedure for using time-charge parking spaces. https://www.navipark1.com/sp/guide/en.html#anc03. Accessed 2 Apr 2020
5. Suzuki, M., Katsumata, Y., Yamada, A.: A study of parking vehicle sensing technology using home wireless LAN equipment. In: IEICE SeMI Conference (2019)
6. Goto, A.: Influence of sense of surveillance by supernatural existence on subjective happiness and social preference. Association of Behavioral Economics and Finance (2017)
7. Ohtsuki, T.: Monitoring techniques with radio waves. IEICE Soc. Mag. **41**, 24–28 (2017)
8. Okugawa, Y., Suzuki, Y., Tajima, K., Yamane, H.: Investigation of human sensing technology uses wireless LAN AP signals. In: IEICE General Conference, B-4-7 (2008)
9. Nishi, M., Kawaguchi, T., Takahashi, S., Yoshida, T.: Proposal on human detection system using UHF band TV receiving wave. IEICE J. B Commun. **J89-B**(9), 1789–1796 (2006)
10. Wang, G., Zou, Y., Zhou, Z., Wu, K., Ni, L.M.: We can hear you with Wi-Fi! IEEE Trans. Mob. Comput. **15**(11), 2907–2920 (2016)
11. IEEE 802.11 WIRELESS LOCAL AREA NETWORKS The Working Group for WLAN Standards. http://www.ieee802.org/11/. Accessed 3 Mar 2020
12. Wire Shark. https://www.wireshark.org/. Accessed 3 Mar 2020

Supporting System for Visiting Sightseeing Spot Using Image Recognition and Processing Technology (KadaBingo)

Soma Ishikawa[1]([✉]), Ryosuke Izumi[1], Hiroki Asakimori[2], Satoru Yamada[2], Takayuki Kunieda[3], Yusuke Kometani[3], Naka Gotoda[3], and Rihito Yaegashi[3]

[1] Graduate School of Engineering, Kagawa University, Hayashi-cho 2217-20, Takamatsu, Kagawa 761-0396, Japan
s20g453@stu.kagawa-u.ac.jp
[2] Ricoh Inc., Nakamagome 1-3-6, Ota-ku, Tokyo 143-8555, Japan
[3] Faculty of Engineering, Kagawa University, Hayashi-cho 2217-20, Takamatsu, Kagawa 761-0396, Japan

Abstract. Utilization of gamification is increasing in the sightseeing filed. Gamification is all activities that uses game design techniques and mechanisms to solve issues and increase customer loyalty. It has been reported that gamification has a certain effect on the increase in the number of visited sightseeing spots and the days of stay in sightseeing spots. In this study, we developed "Supporting System for Visiting Sightseeing Spot Using Image Recognition and Processing Technology (KadaBingo)" and conducted the field experiment in Zentsuzi city, Kagawa Prefecture. This paper describes KadaBingo and analyzes travel behavior based on the data obtained from the field experiment.

Keywords: Sightseeing information · Gamification · Image recognition · Sightseeing behavior analysis

1 Introduction

Maeda [1] defined tourist information as "information which is necessary for tourist in every situation when they sightsee". Yasumura [2] classified tourist information by stage of tourism behavior. Ichikawa [3] describes the classification of sightseeing information made by Yasumura that sightseeing information is classified into three information; "Pre Information" which is necessary at the preparatory stage, "Spot information" which is necessary at the destination, and "Post Information" which is handled after sightseeing, and it is necessary to transmit appropriate information with contents and form according to each stage.

In recent years, gamification [4] has been used in the tourism field. Gamification refers to activities that apply game design and game principles to something other than games. A progress-bar that indicates the progress of the task indicates the target element to the user. It is a mechanism that promotes task progress by visualizing the progress

M. Virvou et al. (Eds.): JCKBSE 2020, LAIS 19, pp. 141–151, 2020.
https://doi.org/10.1007/978-3-030-53949-8_13

of the target element. This is one of the applications of gamification. Stamp-rally and photo-rally in sightseeing spots [5] are efforts to promote tourism using gamification. It has been reported that there is a certain effect on the increase of the number of visiting sightseeing spots and staying time in sightseeing spots.

We have developed "Supporting System for Visiting Sightseeing Spot Using Image Recognition and Processing Technology (KadaBingo)". KadaBingo was developed using gamification to attract tourist to other sightseeing spots. KadaBingo will show tourist photos of the sightseeing spots and bingo cards. The tourist goes sightseeing while searching for the location of the presented photo. When he finds the location of the photo, he takes a photo with the same composition. Then he uploads the photo to the system. If it is determined by the image recognition technology that the images were taken at the same place, one of the squares on the bingo card will be cleared at random. The tourist sightsee in the city with the goal of achieving bingo vertically, horizontally or diagonally. KadaBingo uses a bingo card to form a sphere formation. It guides tourist to sightseeing spots. In addition, it is possible to analyze tourism behavior of tourist using data uploaded by tourist. This paper describes the development of KadaBingo, a demonstration experiment conducted in Zentsuji City, Kagawa Prefecture, and an analysis of tourism behavior using tourist photographs.

2   Development of KadaBingo

This chapter describes "The Development of Supporting System for Visiting Sightseeing Spot Using Image Recognition and Processing Technology (KadaBingo)". Section 2.1 describes the Overview of KadaBingo. Section 2.2 describes the KadaBingo application. Section 2.3 describes the KadaBingo server.

2.1   The Outline of KadaBingo

Figure 1 shows the outline of KadaBingo. KadaBingo consists of "KadaBingo Application" and "KadaBingo Server".

"KadaBingo Application" consists of "Sightseeing Spot Display Function" that presents tourist with pictures of sightseeing spots, "Tourist Guide Function" that guides tourist according to the remaining distance and direction to the target sightseeing spot, "Image Transmission Function" that sends the photographed photos to "KadaBingo Server" and "Bingo Card Display Function" that displays the current bingo card status.

"KadaBingo Server" consists of "Bingo Card Creation Function" that creates a bingo card to be presented to tourist, "Image Recognition Function" that identifies where the images uploaded by the "Image Transmission Function" were taken, "Bingo Judgment Function" that records the current bingo card status and determines whether a bingo exists vertically, horizontally or diagonally, and a "Tourist Information Extraction Function" that extracts tourist information such as the date and time of photography of photos from transmitted photos.

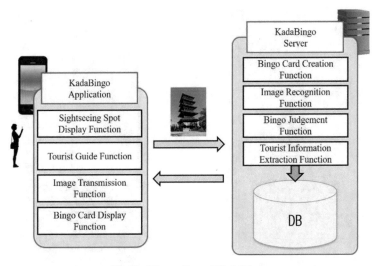

Fig. 1. The outline of KadaBingo

2.2 KadaBingo Application

"KadaBingo Application" is an application to present tourist information and transmit images taken by tourist. It works on mobile information terminals such as smartphones and tablets owned by tourist. "KadaBingo application" consists of a "Sightseeing Spot Display Function", "Tourist Guide Function", "Image Transmission Function", and "Bingo Card Display Function".

"Sightseeing Spot Display Function" is a function that presents a photograph of the sightseeing spots and information of the sightseeing spots to the tourist. Figure 2 shows the screen of the "Sightseeing Spot Display Function". The tourist selects the sightseeing spot that he wants to visit from the displayed sightseeing spot photos. Figure 3 shows information about tourist destinations. By tapping the photo of a sightseeing spot in Fig. 2, the tourist can check information about the sightseeing spot.

"Tourist Guide Function" is a function to guide tourist to a selected sightseeing spot. Figure 4 shows the "Tourist Guide Function" screen. "Tourist Guide Function" presents the remaining distance and direction to the target sightseeing spot to tourist. By tapping the "Photo of destination" button, the photo of the destination is displayed as shown in Fig. 5. The tourist goes sightseeing while looking for places in the pictures based on the remaining distance and direction.

"Image Transmission Function" is a function to send the photograph taken by the tourist to the "KadaBingo Server". The tourist taps the camera icon in Fig. 5 to activate the camera and take a photo, or select a photo saved in the terminal and send it to the "KadaBingo Server".

"Bingo Card Display Function" is a function to display the current state of the bingo card. Figure 6 shows the screen of "Bingo Card Display Function. The tourist can check the current status of bingo cards and the number of bingo. When bingo is achieved, the screen shown in Fig. 7 is displayed.

Fig. 2. Sightseeing Spot Display Function

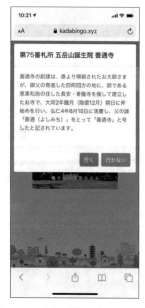

Fig. 3. Sightseeing Spot Display Function

Fig. 4. Tourist Guide Function

Fig. 5. Destination photos

Fig. 6. Bingo Card Display Function   **Fig. 7.** Bingo achievement screen

2.3 KadaBingo Server

"KadaBingo Server" consists of a "Bingo Card Creation Function", "Image Recognition Function", "Bingo Judgment Function", and a "Sightseeing Information Extraction Function".

"Bingo Card Creation Function" creates a bingo card to be presented to tourist. As shown in Fig. 6, a bingo card consists of 3×3 squares. The sequence of numbers displayed on bingo cards is random now.

"Image Recognition Function" is a function that identifies where the image transmitted by the "Image Transmission Function" was taken. "Image Recognition Function" uses Microsoft Azure's Custom Vision [6]. Custom Vision is a cognitive service that allows the user to build, deploy, and improve his own image classifier. An image classifier is an AI service that applies labels (representing "classes") to images based on visual characteristics. In Custom Vision, the user can determine the label to apply. "Image Recognition Function" determines whether or not the image was taken at the sightseeing spot selected by the tourist, and the result of the determination is displayed as shown in Fig. 8 and Fig. 9. In addition, at the time of success, as shown in Fig. 6, one square of the bingo card is randomly cleared.

"Bingo Judgment Function" is a function that records the current state of the bingo card and checks whether the bingo exists vertically, horizontally or diagonally. "Sightseeing Information Extraction Function" is a function to extract EXIF information, which is metadata attached to photos.

2.4 Summary of This Chapter

In this chapter, the development of KadaBingo is described. KadaBingo, developed in this chapter, is a bingo type support system using image recognition/processing technology, KadaBingo presents a photo of a sightseeing spot and a bingo card to a tourist. The tourist search for the location of the photo presented to them while they do their sightseeing. When they find the location of the photo, they take the photo with the same composition. Then, they upload the photos to the system. If the image recognition technology determines that the image was taken at the same location, one of the squares on the bingo card will be cleared at random. The goal of the tourist is to make a bingo iron in one of the vertical and diagonal directions.

KadaBingo consists of a "KadaBingo Application" and a "KadaBingo Server". The "KadaBingo Application" consists of "Sightseeing Spot Display Function" that shows tourist a photo of a tourist location, "Tourist Guide Function" that guides tourist according to the remaining distance and direction to the destination location, "Image Transmission Function" that sends photos taken by tourist to the "KadaBingo Server", and "Bingo Card Display Function" that displays the current status of bingo cards. The "KadaBingo Server" consists of "Bingo Card Creation Function" to create bingo cards to be presented to tourist, "Image Recognition Function" to identify where the image uploaded by the "Image Transmission Function" was taken, "Bingo Judgment Function" to record the status of the current bingo cards and to judge whether bingo exists or not vertical, horizontal or diagonal direction, and "Sightseeing Information Extraction Function" to extract tourism information such as the date and time of shooting from the photos. KadaBingo is able to analyze the tourism behavior of tourist based on the information obtained by the "Sightseeing Information Extraction Function".

Fig. 8. Image recognition success screen   **Fig. 9.** Image recognition failure screen

3 KadaBingo Demonstration Experiment in Zentsuji City, Kagawa Prefecture

This chapter describes a demonstration experiment of KadaBingo in Zentsuji city, Kagawa prefecture, and a tourism behavior analysis based on data obtained from the demonstration experiment. Section 3.1 gives an overview of the demonstration experiment. Section 3.2 describes the analysis of tourism behavior based on the results of the demonstration experiment and the data obtained from the demonstration experiment.

3.1 Outline of Demonstration Experiment

In this study, we conducted a demonstration experiment of KadaBingo in Zentsuji city, Kagawa prefecture to confirm its effectiveness. There are various sightseeing spots in Zentsuji, including Zentsuji, one of the 88 locations in Shikoku. The target sightseeing spots are "Zentsuji Temple", "Zentsuji Sightseeing Exchange Center", "Oubosan Tumulus", "Blackboard Art", "Kumaoka Confectionery Shop", "Miyagawa Noodle Shop", "Old Zentsuji Kaikosha", "JR Zentsuji Station", and "Old Army 11th Division Armory Remains". The demonstration experiment was carried out on 25th and 26th, January, 2020. On both days, several operators were dispatched at the "Zentsuji Temple" and "JR Zentsuji Station", to encourage participation in demonstration experiments and for tourist unfamiliar with ICT.

3.2   Tourism Behavior Analysis Based on the Results of the Demonstration Experiment and Data Obtained from the Demonstration Experiment

Table 1 shows the number of users who used KadaBingo and the number of photos uploaded during the two-day experiment; 558 photos were uploaded by 158 users during the two-day experiment. Figure 10 shows the results of extracting the number of photos taken by time of day. The date and time of the photos acquired from the EXIF information were used for the data collection. Taking photographs is a common activity for tourist. If there is a period of time when a lot of photographs are taken, we can say that tourists are actively engaged in sightseeing during this period of time. From Fig. 10, we can see that tourist started their tourism activities around 9 a.m. and finished them around 5 p.m. The most active hours of sightseeing are between 10 a.m. and 12 p.m. and between 1 p.m. and 3 p.m.

Figure 11 shows the number of photographs taken at sightseeing spots. "Zentsu-ji" was the most photographed, followed by "Kumaoka confectionery store" and "JR Zentsuji station". The number of the "Ouboyama tombs" is 23, and it is the smallest. Since "Zentsuji" is the most visited tourist spot in Zentsuji City and the operator of this site provided information on the use of KadaBingo, the most number of photographs were taken. The "Kumaoka confectionery store" was the closest sightseeing spot from Zentsuji, so many tourists must have visited there.

Figure 12 shows the number of sightseeing spots visited by tourists. The number of users in this demonstration experiment was 158, which means that 158 tourists visited more than one sightseeing spot. Ninety-five of them visited two or more sightseeing spots. Figure 12 shows that most of the tourists who visited the three spots visited the fourth spots. It was also found that there were eight tourists who visited four spots and did not visit the fifth spots, but seven of them had achieved bingo by the fourth spots. In other words, they may have stopped visiting the next sightseeing spot by achieving bingo. Similarly, there were 14 tourists who visited five sites and did not visit the sixth spots, but it was found that 13 of them had achieved bingo by the fifth spots. In the sixth and subsequent locations, all participants achieved one or more bingo when they stopped visiting the sightseeing spots. From the above, it can be assumed that many tourists have increased the number of sightseeing spots they visit in order to achieve bingo.

Table 1. Users using KadaBingo, the number of uploaded photos

	User	Number of uploaded photos
1/25	92	297
1/26	66	261
Total	158	558

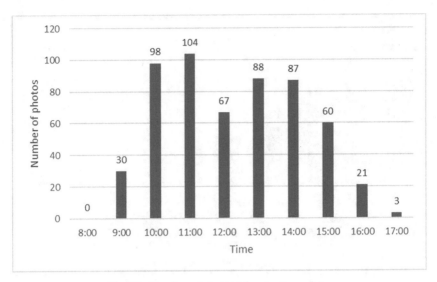

Fig. 10. Number of photos taken by time of day

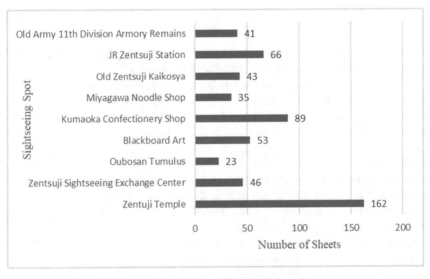

Fig. 11. Number of photos taken by sightseeing spots

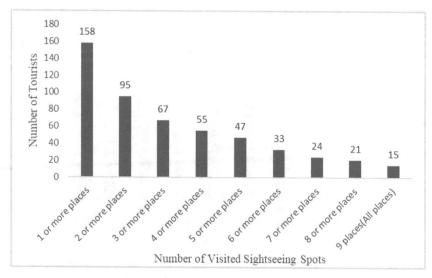

Fig. 12. Number of visited sightseeing spots

4 Conclusion

In this paper, we described "The Supporting System for Visiting Sightseeing Spot Using Image Recognition and Processing Technology (KadaBingo)".

KadaBingo is a bingo-type tourist attraction support system using image recognition/processing technology. KadaBingo presents tourist with a photo of a sightseeing spots and a bingo card. The tourist search for the location of the photo presented to them while they do sightseeing. When they find the location of the photo, they take the photo with the same composition and upload the photo to the system. If the image recognition technology determines that the image was taken at the same location, one of the squares on the bingo card will be cleared at random. The goal of the tourist is to achieve bingo vertically, horizontally or diagonally. KadaBingo is a bingo card based system that guides tourists to sightseeing spots. In addition, it is possible to analyze tourism behavior using photos uploaded by tourists.

We conducted a demonstration experiment of KadaBingo in Zentsuji City, Kagawa Prefecture, Japan, and 558 photos were uploaded by 158 users. As a result of the analysis of the data obtained through the experiment, it was found that the time that tourist actively toured (the time that they took a picture), the time that tourist visited each sightseeing spot (the time that they took a picture at each sightseeing spot), the movement of tourists between sightseeing spots, and the number of sightseeing spots visited by tourist. The results of the analysis of tourism behavior showed that KadaBingo has a certain effect on the number of sightseeing spots and the length of time in those spots. In other words, KadaBingo has the potential to apply gamification as well as stamp-rallies and photo-rallies, to support the formation of a region and to attract tourist to different sightseeing spots in the same region. Thus, KadaBingo has the potential to support statistical data collection in sightseeing spots and contribute to tourism policy decisions.

KadaBingo is a web application built on Azure. In case of deploying the KadaBingo service to other sightseeing spots, it is possible to provide the service to other sightseeing spots by simply increasing the number of terminals without changing the server. By using cloud technology, KadaBingo can flexibly respond to changes in the system, such as horizontal deployment of the system.

Acknowledgment. This research was supported by Ricoh Research Fund, Zentsuji Research Fund, Kagawa University Revitalization Project Expenditure.

References

1. Maeda, I.: Modern Tourism General Remarks, 3nd edn. Gakubunsya (2007)
2. Yasumura, M., Noguchi, Y., Konno, M.: Tourism Business Argument Lecture on Kunpul (2005)
3. Ichikawa, H., Abe, A.: Information technology support on trip around. Jpn. Soc. Artif. Intell. **26**(3), 240–247 (2011)
4. Fukada, K.: It Works If It is a Game. NHK Publishing, Tokyo (2012)
5. Ikeda-ku Tourism Association: Photo Rally. https://www.ikedashi-kanko.jp/media/event/5142#title1
6. MicroSoft: Custom Vision, 01 April 2020. https://azure.microsoft.com/ja-jp/services/cognitiveservices/custom-vision-service//

Development of Support System to Activate Learner's Question Behavior with Dummy Learner

Takumi Shiiki[1]([✉]), Tomoki Yabe[1], Teruhiko Unoki[2], Takayuki Kunieda[3],
Yusuke Kometani[3], Naka Gotoda[3], and Rihito Yaegashi[3]

[1] Graduate School of Engineering, Kagawa University, Hayashi-cho 2217-20, Takamatsu,
Kagawa 761-0396, Japan
s20g466@stu.kagawa-u.ac.jp
[2] Imagica Robot Holdings Inc., Uchisaiwai-Cho Tokyu Bldg. 11F, Uchisaiwaicho 1-3-2,
Chiyoda-ku, Tokyo 100-0011, Japan
[3] Faculty of Engineering and Design, Kagawa University, Hayashi-Cho 2217-20, Takamatsu,
Kagawa 761-0396, Japan

Abstract. In general, learners try to solve problems by questioning behavior when they do not understand. We conducted the survey of learner experiences during the lecture using journey map, persona method and interviews. As a result, it was found that the learner had a psychological burden on question behavior. We have developed a support system to activate learner's question behavior using a chatbot with a dummy learner as a means of reducing the psychological burden. This paper describes the result of the survey of learner experiences during the lecture and the system developed based on the results.

Keywords: Question behavior · Psychological burden · Dummy learners

1 Introduction

Society 5.0 is a system that combines cyber space (virtual space) and physical space (real space) at a high level, and aims to create a anthropocentric society (Society) that balances economic development and solving social issues. DX (Digital Transformation) is attracting attention as one of the means to realize Society 5.0. DX is said to "change people's lives to better ones by penetrating advanced digital technologies". "Digitization" and "Digitalization" can be cited as approaches for DX. "Digitization" refers to partial digitization, such as introducing digital tools to improve efficiency in a certain process. "Digitalization" refers to efforts to digitize the entire process from a long-term perspective, including in-house and external environments and business strategies.

According to the degree of satisfaction with the quality status of system development by year and by system size based on the 2016 corporate IT trend survey report of the Japan Information Systems and Users Association, largescale system development projects of 500 people per months or more, in 2015, 26% of the projects are dissatisfied

© Springer Nature Switzerland AG 2020
M. Virvou et al. (Eds.): JCKBSE 2020, LAIS 19, pp. 152–161, 2020.
https://doi.org/10.1007/978-3-030-53949-8_14

with the system quality. In ISO/IEC25010, as a quality model, "effectiveness", "efficiency", "satisfaction (practicalness, pleasure, comfort)" etc. are separate from "product quality", which refers to the quality of products such as systems and software. Point-in-use quality is defined. It is stated that in order to increase user satisfaction, it is necessary to consider not only product quality but also UX (User Experience), which is a part of in-use quality. UX refers to the total experience gained when users use or consume a product or service, and is a concept that emphasizes whether users can realize what they really want to do in a fun and comfortable way. The method of designing services and systems based on UX from the planning stage of products and services is called UXD (User Experience Design). Currently, this UXD is attracting attention in various industrial fields, and its initiative are expanding. In the field of education, services and systems must be designed in consideration of the learner's experience. Gamification is one of the methods of system development based on UXD. Gamification refers to approach that apply game design and game principles to activities other than games. Fukada [2] describes a framework for applying gamification to actual services and systems as a gamification framework. The gamification framework consists of a total of six elements ("purpose and user", "visualization element", "target element", "social action", "play cycle", and "improvement/operation"). Gamification is also attracting attention from the viewpoint of education and learning support, and Hasegawa describes that gamification has a high affinity with support measures in cognitive skill learning [3]. The use of gamification in systems for education and learning is increasing. In this research, we have developed a question behavior activation support system with the participation of dummy learners to support learning using asynchronous educational video contents. Dummy learners are used to realize a mechanism that activates question behavior by "social action" in gamification. In this paper, we describe the outline of the developed support system to activate learner's question behavior and the result of the demonstration experiment conducted to confirm the effectiveness of the system. The support system to activate learner's question behavior developed in this research not only digitizes teachers and TAs (teaching assistants) and helps them answer effective questions, designed to activate the questioning behavior by consciousness of the learner's experience by the participation of the dummy learner. In other words, the support system to activate learner's question behavior developed in this research is based on UXD, and is a mechanism to improve the quality of the system when it is used in education.

2 Learner Experience Survey in Learning

The purpose of conducting an LX survey is to investigate the learner's experience during the lecture. The Survey contained a persona method, creation of a journey map, and interviews with subjects. The persona setting by the persona method was prepared before the creation of the journey map, and the subjects were interviewed on creating the journey map. In this research, a journey map was created using the persona closest to the user type existing in the environment around the subject. In order to create a journey map with multiple subjects, the work contents and hobbies of part-time workers were set in detail to enhance the common understanding of the personas of the subjects, and consideration was given so that more specific persona images could be imagined. Journey maps are created for the purpose of visualizing not only processes but also user behaviors and emotions in a series of user experiences across multiple touch points. By creating a journey map, the user experience can be shared among stakeholders in terms of time axis. In addition, by showing the whole picture of the user experience, the outline of the user experience considering the points to be improved can be examined. By conducting interviews while comparing the emotion curve with each item in the journey map, the aim is to make the subjects more connected and explain the two items. The interview content was to ask which of the items in the "action" and "thinking/emotional" factors caused the ups and downs of the emotion curve obtained by the survey. The purpose of the interview is to understand the relationship between the psychological situation of the user and the emotion curve that can be considered from the created journey map. Figure 1 shows the emotion curves drawn by three subjects. As a common feature in Fig. 1, it can be seen that the emotion curve swings negatively (negative) in the question generation phase surrounded by the red frame. No clear emotional changes were observed during the question and answer time phase, which is surrounded by the blue frame. Only the emotion curve of subject C slid positively. However, as a result of the interview, Persona's thoughts and behaviors when a question arises, as imaged by Subject C, stated, "I hope that someone else will ask the same questions as I do, without any intention of asking. so a little more positive emotions." As a result, even if the teacher sets the time for question and answer, it can be said that the learner may not intend to use it.

In this survey, a journey map is created for only one type of persona, but there are many reasons why learners cannot ask questions. For this reason, it is considered that the obstacles for learners to ask questions differ from individual to individual. Above all, there were many opinions that concerned the surrounding eyes, such as "I understand the part I did not understand, but I do not have the courage to hear it", "There are parts I care about, but I am ashamed I can not ask questions", "I do not want to know this part if I am the only one", "I do not want to be noticed, I do not want to be angry at the teacher, I do not want to be amazed." It was found that learners hesitate to ask questions due to psychological burdens, even though they were able to clarify questions and verbalize questions by consciousness of others' eyes and evaluations.

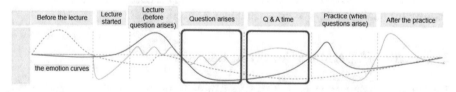

Fig. 1. The emotion curves drawn by three subjects

3 Development of the Support System to Activate Learner's Question Behavior with the Participation of Dummy Learners

In this research, we conducted a demonstration experiment to confirm the effect of dummy learner's participation on learners using the developed support system to activate learner's question behavior using asynchronous educational video contents. Dummy learners in this research were defined as learners who make questions and comments during lectures. Dummy learners are used to realize a mechanism that activates questioning behavior by "social action" in gamification. In this research, we develop a support system to activate learner's question behavior that encourages learners to ask questions using chatbots. Chatbot is a coined word combining "chat" and "bot" which is an abbreviation of robot. Chatbot is a system that can answer questions through simple conversations with users and provide information needed by customers. Chatbot has the characteristic of being able to collect various types of information through conversations, with the "feeling of talking to humans". The survey showed that learners sometimes felt psychologically burdened by being conscious of others' eyes and hesitated to ask questions. In addition, according to the research of Fujii et al., it can be found that the psychological state of the student affects the questioning behavior, and it is possible to activate the class by practicing the class based on them. In this research, we focus on chatbot, which are

Fig. 2. Content distribution function

means other than face-to-face, and develop a support system to activate learner's question behavior. The support system to activate learner's question behavior is a system that supports question behavior, and was developed using the business chat tool "Slack". The support system to activate learner's question behavior is a system that supports learning using asynchronous educational video contents. It is a "content distribution function (function 1)" that distributes educational video content, a "question candidate presentation function (function 2)" that assists in generating question texts related to lectures, and a It has three functions of "question answer function (function 3)" for answering.

The "Content distribution function (function 1)" is a function that distributes educational video content uploaded on YouTube to learners using Slackbot provided by Slack. When a learner inputs a learning item (e.g.,"2-1 Information Encoding") that they want to watch on the Slack channel, Slackbot sends the URL of educational video content according to the learning item. Figure 2 shows the URL transmission scene of educational video contents to the learner using the"Content distribution function".

The "Question candidate presentation function (function 2)" is a function to support learners in generating questions. This system aims to activate learners who are not good at generating questions to ask questions by presenting candidate questions. Figures 3 and 4 show scenes in which a learner is receiving question candidates using the "Question candidate presentation function".

The "Question Answer Function (Function 3)" is a function that answers specific questions (questions prepared in advance for each educational video content) within the Slack channel and answers the questions asked by the dummy learner and the learner,

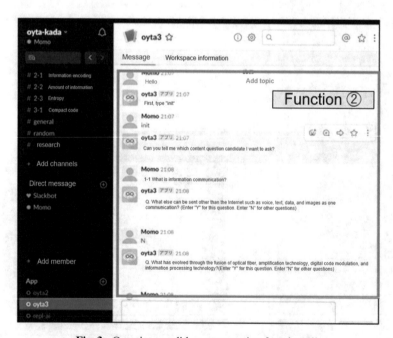

Fig. 3. Question candidate presentation function (1)

and the questions presented by the "Question candidate presentation function". Figures 5 and 6 show the usage scene of the "Question answer function".

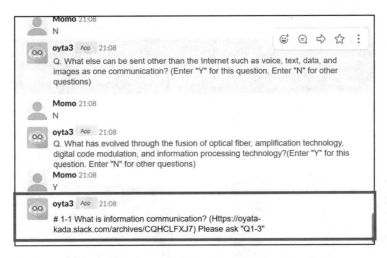

Fig. 4. Question candidate presentation function (2)

4   Demonstration Experiment of Support System to Activate Learner's Question Behavior with the Participation of Dummy Learners

The demonstration experiment was conducted from 15th to 24th January, 2020, with 15 subjects with basic knowledge of information As a test environment, a Slack private channel was prepared for each subject. This experiment proceeded along steps 1 to 6. The procedure in this experiment is shown below.

Step1 Join Slack Private Channel (Experimental Channel)
Step2 Explain the outline of the demonstration experiment
Step3 Watch educational video content (dummy learners perform actions)
Step4 Create customer journey map (create for each content)
Step5 Interview based on journey map
Step6 Questionnaire survey

Subjects asked questions and made comments in the experimental channel. In addition, YouTube was used to view the educational video contents. In step 2, I was told that an explanation of the experiment outline, "the dummy learner is participating in the experimental channel in which the subject participated" and "the dummy learner asks questions and comments on the lesson while watching the educational video content. In addition, as experimental conditions, "do not pause educational video content during learning" and "playback speed of content is basically 1x the standard speed". After

Fig. 5. Question Answer Function (1)   **Fig. 6.** Question Answer Function (2)

the subject watched all three contents, the interview survey in step 5 and the questionnaire survey in step 6 were conducted. The interview survey was conducted to confirm whether the subject's learning motivation was influenced by the dummy learners. In step 6, the effect of dummy learners on question behavior was investigated using a questionnaire. According to the results of the questionnaire, about 70% of the 15 subjects of this experiment (10 of 15 subjects) asked the question. Figure 7 shows the questions for the subjects who performed the questioning behavior in this experiment. According to Fig. 7, all 10 subjects who answered the question answered that the dummy learner "comments" the subject's question as "good" or "slightly good". This shows that all the subjects who asked the questions favorably captured the comments of the dummy learners. Figure 8 shows the results of answering questions about the effectiveness of the question candidate presentation function of the a support system to activate learner's question behavior. According to Fig. 8, 54% of subjects favorably understood the question candidate presentation function of the support system to activate learner's question behavior. On the other hand, 46% of the respondents answered "I didn't feel anything" or "Not very good" about the question candidate presentation function. Reasons why the subjects answered "I did not feel anything particularly" or "Not very good" were "I

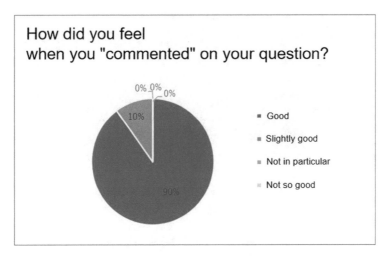

Fig. 7. Questions and answers for the subjects who asked questions in this experiment

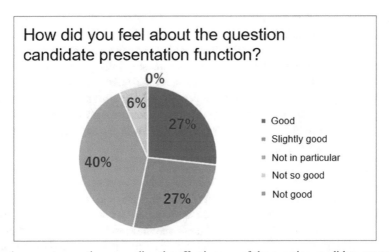

Fig. 8. Answers to questions regarding the effectiveness of the question candidate presentation function

did not use the question candidate presentation function" and "It is difficult to use the question candidate presentation function". For this reason, it became clear that there was a problem with the question candidate presentation function in how learners use it and how to present question candidates. We asked the subjects who did not perform the questioning behavior in this experiment and asked why they did not perform the questioning behavior. As a reason for not asking the question, 4 out of 5 subjects answered that "there was no question for the lecture", and 1 out of 5 subjects answered "I had a question but could not make a question.". Figure 9 shows the answers to questions regarding the effectiveness of the question candidate presentation function. According to the results and

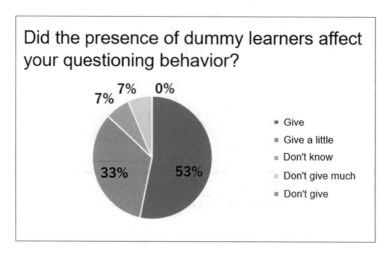

Fig. 9. Answer to the question whether the dummy learner influences the questioning behavior

considerations of a questionnaire conducted to investigate the effect of dummy learners on question behavior after the proof experiment, 13 out of 15 subjects answered "give" and "give little" about the effect of the dummy learner on the questioning behavior. This indicates that the dummy learner may activate the questioning behavior in learning using asynchronous educational video contents. In the free description of the questionnaire, there were many opinions that positively considered the dummy learner. In addition, the subjects who did not perform the questioning behavior also answered, "Since the dummy learners asked the basic questions, there was a role as a supplement to learning."

5   Conclusion

An interview survey based on the journey map and the journey map created in the demonstration experiment showed that the dummy learner affected the motivation of the learner. In addition, the results of the questionnaire survey showed that the learners favored the comments and likes of the dummy learners, and that the dummy learners affected the learners. In the question asking whether the dummy learner had an effect on the questioning behavior, 13 out of 15 subjects answered "give" or "slightly give". They show that there is a possibility to activate the questioning behavior in learning using video contents. In this research, the effectiveness of the support system to activate learner's question behavior by dummy learners' participation was shown. Further studies are needed in order to examine the attributes of the dummy learner and the content and timing of the action of the dummy learner. The support system to activate learner's question behavior developed in this research not only digitizes teachers and TAs to supports effective answers to questions but also construct a mechanism that activates the question behavior by consciousness of the learner's experience by the participation of the dummy learner. In other words, the support system to activate learner's question behavior developed in this research is based on UXD, and is a mechanism to improve the quality of the system when it is used in education.

References

1. Japan Users Association of Information Systems.: Corporate IT Trend Survey Report 2016, Nikkei Business Publications, Inc. (2016)
2. Kouji, F.: It works if it is a game. NHK Publishing, Inc, Japan (2012)
3. Hasegawa, S.: Role of Gamification in Cognitive Skill Learning Support. JPN Soc. Inf. Syst. Educ. **36**, 9–16 (2019)

Deploying Personalized Oblivious Mobile Identity

Georgia Sourla[1,2] and Evangelos Sakkopoulos[3]([✉]) [iD]

[1] Computer Engineering and Informatics Department,
University of Patras, Patras, Greece
[2] Scytales AB, Taby, Sweden
gsourla@scytales.com
[3] Department of Informatics, University of Piraeus, Piraeus, Greece
sakkopul@unipi.gr

Abstract. Mobile Id approach provides a contactless and secure identification process. Standardization towards a common scheme is important to ensure wide-spread adoption through cross-country and cross-region interoperability. Therefore mobile identification has seen enormous attention both by platform vendors and researchers. In this work we discuss a novel mobile identification approach that is based on oblivious identity to ensure improved privacy protection and identification data at rest protection. The mobile identification approach addresses the development of the interfaces to connect users' identities to their real-world physical identities by integrating with officially issued eIDs through bridging with eIDAS, STORK, or electronic driving licenses. In particular we present the initial validation for mobile identification of distributed virtual iDP cryptographic infrastructure for mobile ID use cases. The key aim is to show the particular steps needed in order to perform thorough testing in new cryptographic approaches that aim to be applied in real life application through presented first pilots. Evaluating these prototypes and also evaluating them with other stakeholders in order to identify a wider deployment for business development and exploitation. Initial testing and validation results of the proposed mobile identification approach are encouraging and promising showing strong privacy preservation outcomes.

Keywords: Mobile ID · Digital identity · Oblivious design

1 Introduction

Identification through mobile devices is possible to support secure and safe id verification both in person or in remote based approaches using appropriate architectures. Lots of focus on face to face verification identity has been given

---

The original version of this chapter was revised: Affiliation details of author "Georgia Sourla" have been corrected. The correction to this chapter is available at https://doi.org/10.1007/978-3-030-53949-8_17

in the recent years. Towards the seamless connectivity and communication of mobile id approaches, standardization plays an outstanding role to allow interoperability while ensuring key security measures and risk mitigation approaches. Identity verification is used for several occasions everyday. Proving identity fully or partially is needed to access age controlled products (i.e. tobacco), to pick-up registered mail and parcels, to perform physical access control in facilities and buildings and it can even be applied in most recent flow control events as to allow traffic during COVID-19 restriction measures world-wide [1].

Therefore bringing ID into the mobile has seen enormous attention both by industry and academia using a number of different approaches [10]. In this work, we discuss a novel mobile identification approach that is based on oblivious identity [9] to ensure improved privacy protection and identification data at rest protection [7]. The mobile identification in this work addresses the development of the interfaces to connect users' identities to real-world physical identities by integrating with officially issued eIDs through bridging with eIDAS, STORK, or electronic driving licenses.

This work studies the real working pilots design and approach. The aim is to analyze the possible deployment approaches and the initial working concepts of the mobile solution of corresponding to a mobile identification use case. We include the initial feedback to confirm deployment is aligned to the requirements defined from the real-world situations.

Two possible scenarios were identified, namely an online and an offline scenario. The following sections provide a more detailed explanation regarding the steps that the scenario's flows follow and the expected behaviour for both of them.

In the first scenario, the holder and the verifier are performing the identification process using server based transfer. We call this scenario online, because the holder needs to have at all times connection to the server. The alternative scenario allows the identification to be performed using device to device data transfer using local connections such as BLE, WiFi Aware or NFC. We call this scenario the offline approach.

According to ISO 18013-5 an mobile id (mDL) should be possible to use even if the holder does not have (usually temporarily) connectivity to any online service. For example, that is possible indoor sometimes i.e. in a basement or in an underground parking lots. Furthermore, connectivity issues may happen outdoor while being in the highway or while hiking in the woods.

In the online approach, the holder of the mobile Driver Licence (mDL) must be online during the whole process of both registration and verification. On the other hand, the mDL verifier must be online only during the setup phase, in order to get the public key from the virtual Identity Provider (vIdP) [6]. Then the verification process shall be performed in offline mode, since no other information is needed from the vIdP. From the mDL holder's perspective, we suppose he already possesses a digitally signed mDL on his mobile device (i.e. smartphone, tablet, smart POS), which's authenticity is certified by the corresponding Issuing Authority.

In the offline scenario, the mDL holder is online during the registration process but goes offline during the verification process. The same holds for the mDL verifier, since his device must be online during the setup phase, but the verification process is made in offline mode. From the mDL holder's perspective, we suppose he already possesses a certified mDL on his mobile device [6].

The work below is organized as follows: Sect. 2 includes discussion on related work. Section 3 and Section present the online and offline Scenarios in a step by step structure. Section 5 provides the Software Architecture and solution Integration approach. Section 6 includes the Software Design in high level. Evaluation Plan is discussed in Sect. 7. Section 8 concludes the paper and discusses future steps.

2   Related Work

Privacy Attribute-Based Credentials (P-ABCs) Fig. 1 [9] presents a possible algorithmic approach to achieve pseudo-anonymity and minimum disclosure of data. P-ABCs are based on the similar "offline" architecture as X.509 digital certificates. A key difference between X509 certificates and P-ABC credentials is that the latter provide a tool to derive specialized one-time tokens that only reveal the minimum needed number of pieces of information [9].

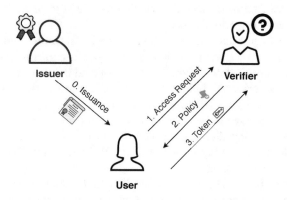

Fig. 1. p-ABC architectural approach

An advanced design for P-ABCs are distributed P-ABCs (dP-ABC). The dP-ABC Credential Management module on the IdP needs to manage the cryptographic material for the scheme. Moreover, it provides the credentials for each user. The analysis of the cryptographic techniques that are being developed for distributed P-ABCs in ObLivious Identity and Management for Private and User-friendly Services are presented further in [3]. ObLivious Identity and Management for Private and User-friendly Services [9] are privacy-preserving identity provision services that use distributed cryptography to distribute the functionality and cryptographic computations of the IdP across multiple authorities, so that no single authority can impersonate or track its users.

Although proposals such as Identity Mixer [8], Distributed Password [7] or European projects such as ARIES [5] represent promising and efficient approaches for the deployment of P-ABC systems, their adoption remains to be seen [4].

In this work, we present the deployment of a solution that incorporates and presents a working approach of ObLivious Identity and Management for Private and User-friendly Services for the use case of mobile Driver's License. The key aim is to protect further aspects of a new interoperable mobile ID (as it is to be specified when the ISO 18013-5 is finalized). One key aspect at hand is the privacy of the backend data stored by the issuing authority. There are more usages that can be further applied such as tracking while being verified in a similar manner.

3   Online and Offline Approaches

According to ISO 18013 specifications, the mDL verifier needs to be online, in order to retrieve the requested data of the mDL holder from the Issuing Authority (IA) itself directly. This a choice by design that aims to facilitate verification without depending on the Internet connection availability on the side of the user. The same model has been used for years in POS based payments. However, in the payment ecosystem the holder has had a passive electronic card that could not connect to the Internet online whatsoever. As a result, in the payments ecosystem, it has been a necessity to have the holders accessing online to perform the necessary transactions. In the mID & mDL ecosystem, the holder and the verifier is almost equally possible to connect online.

More specifically, when the mDL verifier makes a verification request to the mDL holder, then the latter communicates with the IA to get an one-time-only identifier, which grants access to the requested data. The mDL holder sends this identifier to the mDL verifier, so that he will be able to request the needed information from the IA. As long as the mDL verifier can prove that he possesses a valid identifier, the IA shall provide the requested data to the bearer of that identifier. An example of this communication can be seen in Fig. 2 where the merchant of an age restricted good, requests from the mDL holder to prove that his age exceeds a certain limit.

On the contrary, the solution proposed in the Olympus system does not require from the mDL verifier to be online during the verification procedure. In the Olympus approach, the mDL verifier needs to possess the public key that the IA provides for verification. This is the only time needed for the mDL verifier to be online, in order to retrieve the IA's public key. Afterwards, when a verification must be performed, the mDL verifier shall request the needed information from the mDL holder. This time, it is the mDL holder's responsibility to provide a token to the mDL verifier, which, combined with the IA's public key, can verify the mDL holder. This token can be generated either by the IA upon request, when the mDL holder is online, or by the mDL holder application itself when it is offline, as long as it already possesses certified credentials from the IA, stored in the holder's device. The same example for the online case can be viewed in Fig. 3, in order to show the differences between the ISO and the Olympus solution.

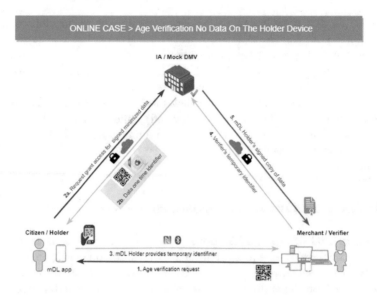

Fig. 2. Online flow according to ISO 18013-5

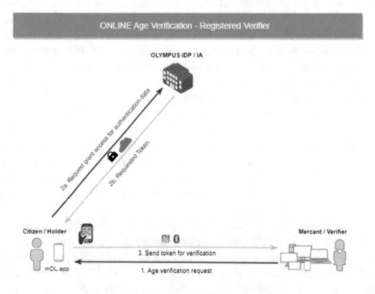

Fig. 3. Online flow in Olympus solution

4   Online Distributed Token Approach

The workflow from the mDL holder's and mDL verifier's perspective can be described using the following step-by-step approach in the online case.

VERIFIER SETUP:
For setting up the mDL verifier application, the verifier needs to start it and press the setup button, in order to obtain the public key from the vIdP server. Once the key is successfully received, the mDL verifier application is ready to perform verifications.

VERIFICATION:
A verification procedure may start when the mDL verifier initiates a connection with the mDL holder's device, in order to send the policy that the holder needs to fulfill. When the policy is received by the mDL holder application, the mDL holder will see which attributes shall be revealed to the mDL verifier. If the mDL holder chooses to accept the policy, a request is performed from the mDL holder application to the vIdP, in order to provide the mDL holder with a token for the specified policy. After receiving the token, the mDL holder application forwards the token to the mDL verifier, so that he can proceed with the verification process. Once the token is sent from the mDL holder application, the mDL verifier can verify it against the public key obtained during setup. In the end, a message about the result of the verification appears on the mDL verifier application screen.

The main advantage of this approach is that during every verification, the most updated identification attributes are used in order to generate a token, so the mDL verifier can be sure that he verifies the mDL holder against updated personal information. On the other hand, this approach requires that the mDL holder has an online connection established throughout the whole verification procedure.

5 Offline Scenario dP-ABC Credentials

The offline case is quite different, since only the setup phase requires an online connection. As a result, the mDL verifier application is setup with the same steps, described in the previous section. The verification procedure though is now performed following the process described below.

VERIFICATION:
In an offline verification, the mDL verifier initiates a connection with the mDL holder's device, in order to send the policy that the holder needs to fulfill. When the policy is received by the mDL holder application, the mDL holder will see which attributes shall be revealed to the mDL verifier. Up to this point, the procedure follows the same steps as in the online scenario. This time though, if the mDL holder chooses to accept the policy, it is the mDL holder application and not the vIdP that shall generate a token based on the credentials acquired from enrolment procedure and the policy received from the verifier. After generating the token, the mDL holder application forwards it to the mDL verifier, in order to proceed with the verification process. The mDL verifier receives the token from the mDL holder application and verifies it against the public key obtained during setup. As in the online scenario, the verifier needs no further communication

with the vIdP, after acquiring the public key, so the whole verification procedure is performed in offline mode. In the end, a message about the result of the verification appears on the mDL verifier application screen.

With this approach, there is no need for the mDL holder to be connected during verification, thus he can be verified and make use of the requested service at any time needed. On the other hand, the mDL verifier can perform verification based on the personal information that is currently stored in the mDL holder's device. We expect though, that the mDL holder updates his info in a regular base, in order to avoid expiration.

6   Differences in Online and Offline Scenarios

The Online and Offline scenarios in the mDL use case are prescribed in the current draft of ISO 18013-5. Having in mind the physical driver's license, passport and id verification processes, it is mainly needed to cover that any electronic based identification will be possible always independent of any long rang connectivity such as reaching a server based online service or piece of information.

For this reason, the offline scenario dictates that the verification environment between an mDL holder and an mDL verifier is only the transfer of data from the holder's device to the verifier's locally with no server connection. Such connectivity can be achieved in an optical manner using a QR-code or by tapping devices that are near field communication (NFC) enabled. Next the devices transfer data using short range transmission technologies such as Bluetooth Low Energy (BLE), WiFi Aware or NFC. All these connectivity options are popular in smartphones and at least one practically exists in any of them.

Alternatively, in the mDL use case it is possible to connect to a server of any issuing authority to perform online verification. The online scenario for mDL is similar to online payments using a plastic card and a POS device when it comes to necessary availability for wide area network connectivity.

Section 3 and 4 describe the steps that have been devised to perform verification under each scenario. The key difference is in the setup phase that requires a server based connection. In order to facilitate the needs of the offline scenario the holder's device is communicating with the identity services in an ad hoc manner but in advance of any identification and verification process. The mDL holder is keeping the necessary information on the device after that is downloaded from the server for a short term duration. The duration can be the typical shift duration (4–8 h) or one of a few days.

7   Architecture and Integration

The following sections describe how the integration was performed between each component and the current state overall.

7.1 Interfaces

The mDL applications for both holder and verifier need to communicate with the vIdP. From the holder's side, there is the need to create an account to the vIdP which includes the holder's personal information, that would be available for authentication, depending on the verification requirements. There is also the need for token creation, that would be generated either directly from the vIdP in the online scenario or from the mDL holder in the offline scenario. The latter is possible by utilizing the dP-ABC Client component, which can store the mDL holder's credentials, obtained from the vIdP during enrolment, and use them anytime later in order to create tokens for the specified policies in offline mode. The interfaces used for the aforementioned procedures can be viewed in Fig. 4 and Fig. 5.

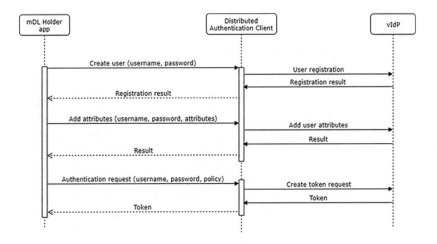

Fig. 4. Interfaces for mDL Holder (online scenario)

From the verifier's side, either for the online or for the offline scenario, the only connection needed to be established with the vIdP, is made on the setup phase for acquiring the public key that shall be used during verification.

7.2 Components

From the previous section, we can easily understand that there are quite a number of modules being used for the mDL use case. The two basic ones for this testing scenario are the mDL holder application and the mDL verifier application. These two applications allow us to demonstrate the mDL use case by pretending to have a real mDL holder that is being verified against a certain policy (i.e. proof that the holder's age is above a certain limit) by an mDL verifier. Currently two different systems are supporting the online and the offline scenario.

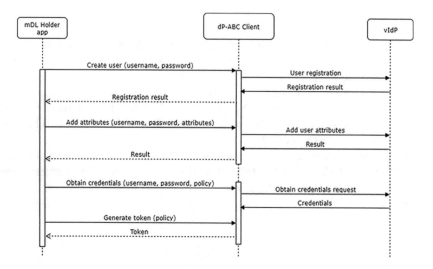

Fig. 5. Interfaces for mDL Holder (offline scenario)

In the online scenario, the Distributed Authentication Client module is utilized in order to perform registration and authorization. This module establishes a connection with the vIdP server in order to create a new user account, store the mDL holder's personal information and provide to the holder an authorization token compliant to a specific policy. In the offline scenario, this part is played by the dP-ABC Client component. The main difference between these two modules is that the dP-ABC Client component needs to store the mDL holder's P-ABC credentials, in order to be able to generate tokens in offline mode at a later point in time. On the contrary, the Distributed Authentication Client module does not need to store information about the mDL holder and directly forwards the generated token from the vIdP to the mDL holder application.

8   Design and Development of Components

Currently, for the mDL use case we have been focusing on the offline scenario for the first pilot deployment. We are expecting though that dealing only with the offline scenario will pay back when we turn to the online one, not only because of the feedback that would be received from the first demo testing, but also from all the issues that already have been overcome through this initial offline implementation of the mDL use case.

8.1   The mDL Issuing Authority

The mDL Issuing authority role is performed by an instance of the vIdP.

As an initial vIdP pilot deployment process for the offline scenario, three instances of the Proactively Secure Distributed Single Sign-On (PESTO) IdP were deployed as an embedded servlet container. These three instances would provide distribution during all mDL holder's actions, such as enrollment, adding identification attributes and issuing credentials. Additionally, all instances are aware of the existence of the rest IdP instances and each one of them is configured with a JSON file that is used during each PESTO IdP startup phase.

For the offline scenario pilot, the mDL holder application connects to the three IdPs via dP-ABC client interface, while the mDL verifier application utilizes the dP-ABC verifier interface in order to connect to the vIdP. These two interfaces provide all the needed functionalities from a single point, and as a result both mDL applications are able to perform client registration, identity attachment, credential issuance and token verification, accordingly.

8.2 The mDL Apps

For the offline mDL scenario, two applications have been implemented, one for the mDL holder and one for the mDL verifier. They are implemented in such a way that the mDL holder application initiates the mDL verifier application while sending the authentication token.

With the mDL holder application, the holder is able to create his own account at the vIdP, by registering with a username and password of his choice. The implemented application gathers the typed credentials and the mDL holder's personal information from the mDL he possesses and sends all needed data to the vIdP, in order to create an account and store the holder's personal info. The result of this procedure is shown to the mDL holder with an appropriate message.

After a successful registration, the mDL holder can request to obtain his P-ABC credentials from the vIdP and keep them for later use. This is the last step where the mDL holder needs to be online. From now on, he is able to generate an authentication token for the policy supported by the application, without being connected to the vIdP. When the token is generated successfully, the mDL holder app shall initiate the mDL verifier app, providing the acquired token. As in the registration procedure, credentials are not currently stored in the holder's device.

During the setup phase, the mDL verifier application is connected to the vIdP, in order to acquire the public key needed for the verification procedure. Like the mDL holder application, the public key is not stored in the device, so the setup phase should be performed again the next time the application starts. With the public key acquired from the vIdP and the token sent from the holder's application, the verifier can continue with the verification step. Depending on the result of this function, an appropriate message shall be printed in the verifier's screen.

9 Evaluation

The following sections describe how the demonstrations that took place for the offline scenario should be replicated, as well as the testing plan for the mDL use case. In the experimental prototype section functionality testing is presented to show that the basic necessary mDL use case functions are covered focusing on mDL initialization (also called sign-up) and the verification process for valid and invalid cases. Next, unit testing and integration testing is described, where the mDL has been interconnected and adapted to use distributed and oblivious approach. We have verified that the all modules are communicating successfully and in a valid manner.

9.1   Functional Verification of the Experimental Prototype

For demonstrating the mDL use case, two Android mobile applications were developed, one for the mDL holder and one for the mDL verifier. The mDL holder application needs to be online during the enrollment phase. We assume that the holder already possesses a digitally signed mDL on his device. After a successful registration, the mDL holder application presents the holder's personal information on the screen. The holder is able to request his P-ABC credentials from the vIdP and go offline (Fig. 6). Now he can generate at any time needed the authentication token which fulfils the policy that the mDL holder application supports. This policy is checking the name and the age of the mDL holder for a standardized age verification process.

Fig. 6. mDL Holder app - credentials request and result

The mDL verifier application needs to be online only during the setup phase. When the verifier starts the verifier application, he just needs to press the "Setup" button, in order to acquire the public key from the vIdP and prepare his application for verification procedure. After that, he can go offline and wait for a token to be received.

In order for the holder to be verified, he needs to press the "Send Token" button in the mDL holder application, in order to generate a token for the supported policy and forward it to the verifier. After this action, the mDL verifier application appears and the verifier may now press the "Verify" button. The verification procedure shall start, in order to check the name and the age of the mDL holder and the result will appear on the screen with a proper message (Fig. 7).

Fig. 7. mDL Verifier app - successful age and name verification

Within demo scope, there is one more button added in the mDL holder application that simulates the case of a fraud holder. What this button does is to alternate the signature of the P-ABC credentials received from the vIdP before generating the authentication token. If this button is pressed during verification process, the mDL verifier application shall show a message of failed verification (Fig. 8).

9.2   Unit and Integration Testing Plans

Currently the testing plan for the mDL use case involves unit and integration testing. As a first step, unit tests were created in order to test small parts of implementation for the two mDL applications under development. At the moment, we have reached a stage where servers are hosting the vIdP integrated system and the mDL applications are connected to them in order to perform the registration and verification procedures.

Unit tests were developed in two phases. At first, simple parts of the implementation were developed along with the user interface, in order to have a basic skeleton for our applications. During this phase, no communication was supported with external systems. On a next level, a first stable version of the new cryptographic system proposed in OLYMPUS was available, with support of dP-ABC credentials. The whole process of registration and verification was tested, with the cryptographic modules being set up in a local server. The same process was also followed when the whole Olympus integrated vIdP system was hosted in one partner's server for the demo tests.

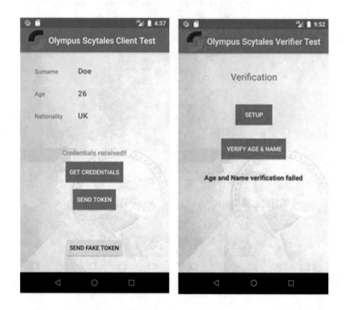

Fig. 8. mDL Holder and Verifier app - failed age and name authentication (b)

In all the testing steps mentioned above, both successful and failed scenarios were tested, so that the integrity of not only the two developed mobile applications, but also of the whole Olympus system would be tested. The results of unit and integration tests have been positive and promising.

9.3   User Survey

Surveys are used to identify the users' views and feedbacks to understand their expectations for a given product, using questionnaires and interviews. In order to evaluate the use case scenario of using a mobile id based on Driver's Licence, we have conducted an initial hands-on user evaluation study during 2.2020–4.2020 using polls of questionnaires from 17 users of age 25–55 years old, to measure the impact of an upcoming mobile id, the expected adoption and usage rate.

The findings show that users show that their intent to use a mobile ID amidst COVID-19 [1] is largely immediate (see Fig. 9) as it gives them the capability to prove age and identity in a contact-less manner, either through QRcode or NFC Based verification. In particular, the users are mainly fond of the privacy preservation features that we show in our solution (See Fig. 10).

Furthermore, responses show that there are several use cases that may be served with a mobile ID such as age verification, picking up a parcel, travelling identification and online site access if possible. For all these services the users expect to use an mobile ID app at least once a month and more (See Fig. 11).

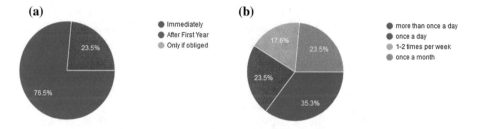

Fig. 9. (a) Intention to use and (b) Frequency of usage

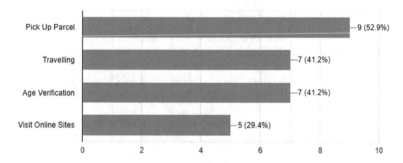

Fig. 10. Services to use mobile ID

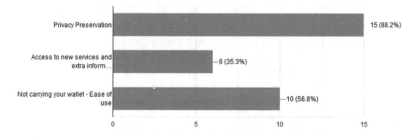

Fig. 11. Mobile ID brings new functionalities

10   Conclusions and Future Work

As we reached the middle of the timeline of the project, there has been a lot of work being done for the development of not only the new cryptographic systems proposed in this project, but also the demo pilots that shall be used for the testing of the aforementioned systems. Yet, there are further steps to be done for having a fully functional integration of the mDL pilot with the rest of the system, whose development is still ongoing.

A further addition would be to securely store data in the devices of the mDL holder and the mDL verifier. With this extension, the two mobile applications shall have all required data, which were already acquired, available for later usage.

Last but not least, the online version of distributed token shall also be supported in the mDL pilot at a later time. With this extension, the mDL holder will be able to get online the token that complies with the policy specified by the mDL verifier he encountered and pass it straight ahead to the verifier's device in order to proceed with the verification.

Acknowledgment. The authors of this paper would like to thank the University of Piraeus Research Center for the financial support of this research paper. This approach has also received funding from the European Union's Horizon 2020 research and innovation program under grant agreement No 786725 (OLYMPUS project).

References

1. Covid-19 restrictions. FRONTEX: European Border and CoastGuard Agency (2020). https://frontex.europa.eu/media-centre/news-release/covid-19-restrictions-4IdY3J

2. Agrawal, S., Miao, P., Mohassel, P., Mukherjee, P.: PASTA: password-based threshold authentication. In: Proceedings of the 2018 ACM SIGSAC Conference on Computer and Communications Security, CCS 2018, Toronto, ON, Canada, 15-19 October 2018, pp. 2042–2059 (2018). https://doi.org/10.1145/3243734.3243839

3. Baum, C., Frederiksen, T.K., Hesse, J., Lehmann, A., Yanai, A.: PESTO: proactively secure distributed single sign-on, or how to trust a hacked server. IACR Cryptology ePrint Archive, p. 1470 (2019). https://eprint.iacr.org/2019/1470

4. Bernabé, J.B., Cánovas, J.L., Ramos, J.L.H., Moreno, R.T., Skarmeta, A.F.: Privacy-preserving solutions for blockchain: review and challenges. IEEE Access **7**, 164908–164940 (2019). https://doi.org/10.1109/ACCESS.2019.2950872

5. Bernabé, J.B., David, M., Moreno, R.T., Cordero, J.P., Bahloul, S., Skarmeta, A.F.: ARIES: evaluation of a reliable and privacy-preserving european identity management framework. Future Gener. Comput. Syst. **102**, 409–425 (2020). https://doi.org/10.1016/j.future.2019.08.017

6. Camenisch, J., Drijvers, M., Lehmann, A., Neven, G., Towa, P.: Short threshold dynamic group signatures. IACR Cryptology ePrint Archive **2020**, 16 (2020). https://eprint.iacr.org/2020/016

7. Camenisch, J., Lehmann, A., Neven, G.: Optimal distributed password verification. In: Proceedings of the 22nd ACM SIGSAC Conference on Computer and Communications Security, Denver, CO, USA, 12-16 October 2015, pp. 182–194 (2015). https://doi.org/10.1145/2810103.2813722

8. Camenisch, J., Mödersheim, S., Sommer, D.: A formal model of identity mixer. In: Kowalewski, S., Roveri, M. (eds.) Formal Methods for Industrial Critical Systems - 15th International Workshop, FMICS 2010, Antwerp, Belgium, 20-21 September 2010. Proceedings. Lecture Notes in Computer Science, vol. 6371, pp. 198–214. Springer (2010). https://doi.org/10.1007/978-3-642-15898-8_13
9. Moreno, R.T., Bernabé, J.B., Rodríguez, J.G., Frederiksen, T.K., Stausholm, M., Martínez, N., Sakkopoulos, E., Ponte, N., Skarmeta, A.F.: The OLYMPUS architecture - oblivious identity management for private user-friendly services. Sensors **20**(3), 945 (2020). https://doi.org/10.3390/s20030945
10. Sakkopoulos, E., Ioannou, Z., Viennas, E.: Mobile personal information exchange over BLE. In: 9th International Conference on Information, Intelligence, Systems and Applications, IISA 2018, Zakynthos, Greece, 23-25 July 2018, pp. 1–8. IEEE Computer Society (2018). https://doi.org/10.1109/IISA.2018.8633599
11. Sanchez, J.L.C., Bernabé, J.B., Skarmeta, A.F.: Integration of anonymous credential systems in IoT constrained environments. IEEE Access **6**, 4767–4778 (2018). https://doi.org/10.1109/ACCESS.2017.2788464

How the Apriori Algorithm Can Help to Find Semantic Duplicates in Ontology

Irina Astrova[1]($\boxtimes$), Arne Koschel[2], and Su Ling Lee[2]

[1] Department of Software Science, School of IT, Tallinn University of Technology, Akadeemia tee 21, 12618 Tallinn, Estonia
irina@cs.ioc.ee
[2] Faculty IV, Department of Computer Science, University of Applied Sciences and Arts Hannover, Ricklinger Stadtweg 120, 30459 Hannover, Germany
akoschel@acm.org

Abstract. Ontology-based data integration attempts to overcome the semantic heterogeneity problem in data integration. Semantic heterogeneity refers to an ambiguous interpretation of terms that describes the meaning of data in heterogeneous resources. However, the presence of semantic duplicates such as similar attributes in the integrated ontologies can lead to incomplete query results. This paper proposes to use the Apriori algorithm from market basket analysis to find similar attributes in an ontology.

Keywords: Ontolgy · Semantic heterogeneity · Market basket analysis · Apriori algorithm · Similar attributes · OntoLife

1 Introduction

The main purpose of ontology-based data integration is to solve the semantic heterogeneity problem in data integration [18]. An ontology is a set of assertions that are used to model a given domain and to define a vocabulary used by the underlying application. Semantic heterogeneity refers to the ambiguous interpretation of similar terms, which are being used to describe the meaning of data in heterogeneous resources such as websites or databases. However, during an ontology-based data integration, the newly merged ontology can contain semantic duplicates, e.g., similar attributes [19]. These attributes can cause incomplete query results.

As an example, consider a user who submits the following query against the Wikipedia ontology: *"Which performers were born in Chicago?"* In response to this query, the query-answering system will return only one result (viz. Michael Ian Black). However, if it were known that *actor* and *comedian* are subclasses of *performer* and that their attributes *birthplace*, *birth place*, *city of birth*, *place of birth* and *origin* are semantic duplicates of performer's location, the query-answering system could return 163 additional results [1]. Thus, the recall of query results can be greatly improved by finding similar attributes.

© Springer Nature Switzerland AG 2020
M. Virvou et al. (Eds.): JCKBSE 2020, LAIS 19, pp. 178–191, 2020.
https://doi.org/10.1007/978-3-030-53949-8_16

In this paper, we propose to use market basket analysis to find similar attributes. In market basket analysis, two items (or products) are considered as similar if the purchase patterns of their customers are similar. Analogously, two attributes are considered as similar if the querying patterns of their users are similar [2]. Continuing the example with the Wikipedia ontology: If it were known that there are many users who have asked about the actor's *birth place* together with the actor's *name* and *birth date*, and that there are many users who have asked about the actor's *origin* together with the actor's *name* and *birth date*, we could conclude that attributes *birth place* and *origin* in a class *actor* are similar to each other.

2 Market Basket Analysis

Baskets are defined as sets of products bought together by customers in sales transactions during their visits to supermarkets or online shops. Market basket analysis examines the buying behaviour of customers and it is used to identify similar products based on the purchase patterns. A typical example of similar products are Becks and Heineken beers. At the first sight, these two products may appear dissimilar as they do not have many common customers: the customers buy either Becks or Heineken, but not both. However, it was observed that together with Becks and Heineken, many of the customers buy the same products such as chips, sausages and pretzels. Based on this observation, a similarity measure is defined as two products are found to be similar, if the purchase patterns of their customers are similar.

The aim of market basket analysis is that of finding sets of items (products) that appear together in (are related to) many baskets. Let products be attributes (search terms in queries), baskets be queries and the sales transaction database be the query log. In this application of market basket analysis, we look for sets of attributes that appear together in many queries.

In this paper, we use the Apriori algorithm [3] from market basket analysis to find similar attributes in the OntoLife ontology. This algorithm analyses many-to-many relationships between the querying patterns and the user behaviours. Users who have similar questions in mind tend to submit similar queries, which have similar querying patterns. Therefore, the Apriori algorithm uses the query log, which contains a list of queries posed by users to inquire against the ontology. With the help of this algorithm, we derive the querying patterns to find similar attributes in an ontology.

While the subject of the Apriori algorithm is extracting frequent sets of items from data, this information is often presented as a collection of "if–then" rules called association rules. For example, if customers buy Heineken, then they also buy chips, sausages and pretzels. Therefore, the Apriori algorithm goes through two stages: (1) finding frequent itemsets; and (2) generating strong association rules from those frequent itemsets (see Fig. 1).

Algorithm 1 Apriori Algorithm - Pseudo code

Algorithm Apriori (D, minsup)
L_1 = {large 1-itemsets};
for ($k = 2$; L_{k-1} is not empty ; $k++$) **do**
    C_k = APRIORI-GEN(L_{k-1}) // generate new candidates itemsets
    // count support
    for all transactions $t \in D$ **do**
        C_t = subset(C_k , t); // candidates itemsets contain in transaction t
        for all candidates $c \in C_t$ **do**
            $c.count++$;
        end for
    end for
    L_k = { $c \in C_k \mid c.count \geq minsup$ }
end for
Output: $\cup_k L_k$;

Function: APRIORI-GEN(L_{k-1})
insert into C_k //join step
select $p.item_1$, $p.item_2$, ..., $p.item_{k-1}$, $q.item_{k-1}$
from L_{k-1} p, L_{k-1} q
where $p.item_1 = q.item_1$,..., $p.item_{k-2} = q.item_{k-2}$, $p.item_{k-1} <$ $q.item_{k-1}$;
// prune step
for all itemsets $c \in C_k$ **do**
    for all (k-1)-subsets s of c **do**
        if ($s \notin L_{k-1}$) **then**
            delete c from C_k;
        end if
    end for
end for
Output: C_k

Function: GENRULES(L_k, minconf) // to generate association rules
for each L_k **do**
    generate its non-empty subsets s;
end for
for every non-empty subset s of f **do**
    if (support(f) / support(s) > minconf) **then**
        output association rule $s \Rightarrow (f - s)$
    end if
end for
Output: Set of association rules for a frequent k-itemsets f

Fig. 1. Apriori algorithm [3].

3   OntoLife

OntoLife [4] is an ontology for semantically managing personal information. This ontology has 110 attributes called datatype properties in the OWL terminology. It was developed by Intelligent System and Knowledge Processing Research Group (ISKP) by merging 10 other ontologies:

1. **Person** [5]: This ontology provides a formal vocabulary description of people. Furthermore, it defines constraints (e.g., a person should have exactly one name). It does not have any attributes.
2. **Friend of a Friend (FOAF)** [6]: This ontology describes the people's activities and the relationships such as friendships between them. It does not have any attributes.
3. **Family Tree** [7]: This ontology models the application domain of family that defines people's relationships such as children and parents. It does not have any attributes.
4. **Relationship** [8]: This ontology defines other relationships between people such as friends, colleagues and tutors. It does not have any attributes.
5. **ISO lists for Countries and Languages** [9]: This ontology models the application domain of people's countries and their languages. It does not have any attributes.
6. **Project** [10]: This ontology describes research projects and people who are involved into those projects. It has 6 attributes such as project title, start and end dates.
7. **Research** [11]: This ontology describes the hierarchy of research projects (e.g., related projects can include cloud computing and data mining). It does not have any attributes.
8. **Publication** [12]: This ontology models the application domain of people's publications. It has 27 attributes such as book title, edition of publication and publication date.
9. **PersonProjectAssociation** [13]: This ontology models the relationships between people and publications or research projects. Furthermore, it defines constraints (e.g., a person should have at least one role in a research project). It does not have any attributes.
10. **Biography** [14]: This is the largest ontology. It has 77 attributes. It describes the people's demographics, their profiles and contacts. Furthermore, it describes the CV-related information such as people's education and qualifications, skills and working experiences as well as injuries that people have got at work.

We selected OntoLife for our experiments because it is from an easy-to-understand application domain. Furthermore, the ontology has many attributes and is merged from many other ontologies from the same application domain. Thus, it is likely to have similar attributes.

4 Experiments

To run the Apriori algorithm against the OntoLife ontology, we made a number of assumptions about how data are stored and manipulated when searching for frequent itemsets (similar attributes):

- In market basket analysis, a basket contains only a small set of products from all products available in the supermarket or online shop. Similarly, it is also assumed that users do not inquire about all attributes in the ontology at once in a single query.
- In market basket analysis, a basket can contain any number of products. Similarly, it is also assumed that a query can contain any number of attributes (i.e., the query can be of any size).
- In market basket analysis, the details on sales transactions such as the date of purchase, product price and product quantities are ignored. Similarly, the values of attributes in the ontology are ignored.
- In market basket analysis, the order of products in a sales transaction is ignored. Similarly, the order of attributes in a query is ignored.
- In market basket analysis, sales transactions are mined for a long period of time (at least for 6 months). As a result, the sales transaction database is large. Similarly, it is also assumed that the log of queries issued against the ontology is large.
- All queries are taken into account, regardless of whether they return results or not.

The query log is the result of the past user interactions with OntoLife. Figure 2 shows an excerpt of the query log, which comprises 30 queries. These queries are written in SPARQL and contain 41 attributes, which constitute approximately 37% of the total attributes in the ontology. The shortest query has one attribute, whereas the longest query has 5 attributes. Not only can queries be distinguished by their size, but also by their frequency occurrence in the query log. A unique query appears only once in the query log. An example of a unique query is: *"Who have "rain man" as their nickname?"* In contrast, repeated queries are the queries, which appear more than once in the query log. They queried on the same attributes but possibly with different values or in a different order. An example of repeated queries is: *"Who were injured on November 22, 2009, were 50% disabled by that injury and worked in the public sector?"* and *"Who were injured on January 2, 1989, worked in the government sector and were 20% disabled by that injury?"*.

To be used as the input for the Apriori algorithm, the query log was converted into a sales transaction database, which contains 30 sales transactions (one sales transaction per query). Figure 3 shows an excerpt of the sales transaction database. In this database, each query is identified by TID (transaction identifier). For each row, if an attribute does not appear in the query, it will not appear in the corresponding row either.

The Apriori algorithm finds frequent itemsets, by combining all possible itemsets with each other and pruning those whose support is lower than a user-defined minimum support threshold. After that, the Apriori algorithm generates strong association rules from the frequent itemsets, by combining them with each of their subsets and pruning those whose confidence is lower than a user-defined minimum confidence threshold. We conducted a number of experiments with different values for the minimum support

Nr.	Query in SPARQL:

01	PREFIX bio:≺"http://users.auth.gr/ elkar/thesis/Biography.owl#" ≻
	SELECT ?name
	WHERE {
	?person bio:awardTitle ?awardTitle .
	?person bio:awardDate ?awardDate .
	?person bio:shipmentDate ?shipmentDate.
	?person bio:internationalName ?internationalName .
	FILTER ((?awardTitle = "Best Bookseller") ||
	(?awardTitle="Oddest Booktitle Prize")
	&& (?awardDate = 2012-02-24)
	& (?shipmentDate = 10-Sept-2012) &&
	(?internationalName = "Springer"))
	}

02	PREFIX bio: ≺"http://users.auth.gr/ elkar/thesis/Biography.owl#" ≻
	SELECT *
	WHERE {
	?person bio:disab_type ?disab_type .
	?person bio:weight ?weight .
	FILTER ((?disab_type = "persistent delusion" &&
	?disab_type = "depression") && ?weight ≻ 80)
	}

03	PREFIX bio: ≺"http://users.auth.gr/ elkar/thesis/Biography.owl#" ≻
	SELECT ?name, ?nickname
	WHERE {
	?person bio:nickname ?nickname .
	FILTER (?nickname = "rain man")
	}

04	PREFIX bio: ≺"http://users.auth.gr/ elkar/thesis/Biography.owl#" ≻ SELECT ?name, ?nationality WHERE { ?person bio:nationality ?nationality . ?person bio:dateofBirth ?dateofBirth . ?person bio:militaryUnit ?militaryUnit . ?person bio:militaryGrade ?militaryGrade . FILTER (?nationality = "French" && ?dateofBirth = "2000-03-29" && militaryUnit = "Eagle" && militaryGrade = "Corporal")

Fig. 2. Query log for OntoLife.

05	PREFIX bio:≺"http://users.auth.gr/ elkar/thesis/Biography.owl#" ≻ SELECT ?name WHERE { ?person bio:name ?name . ?person bio:height ?height . ?person bio:disab_type ?disab_type . ?person bio:weight? weight . ?person bio:selfGrade? selfGrade . ?person bio:eyecolor? eyecolor . FILTER (?height ≺ "1.0 metres" && weight ≻= 80 kg && (eyecolor = "hazel" || eyecolor = "blue") && disab_type = "depression" && selfGrade = "good") } ORDER BY ?name
06	PREFIX bio: ≺"http://users.auth.gr/ elkar/thesis/Biography.owl#" ≻ SELECT ?name WHERE { ?person bio:name ?name . ?person bio:journal ?journal . ?person bio:note ?note . FILTER ((?journal = "Newsweeks") || (?journal="Times") && note = "Fifty cents") } ORDER BY ?name
....	....
....	....

30	PREFIX bio: ≺"http://users.auth.gr/ elkar/thesis/Biography.owl#" ≻
	SELECT *
	WHERE {
	?person bio:nickname ?nickname .
	?person bio:specificDate ?specificDate .
	?person bio:disab_percentage ?disab_percentage .
	?person bio: hasGovernmentalInvolvement
	?hasGovernmentalInvolvement .
	?person bio:sector ?sector .
	FILTER ((?specificDate = "2009-11-23" && sector ="private non-
	business" && disab_percentage = 10 && hasGovernmentalInvolvement
	= 0 && nickname = "flowerpot"))
	}

Fig. 2. (*continued*)

TID:	Items:
001	{awardTitle, awardDate, shipmentDate, internationalName}
002	{disab_type, weight}
003	{nickname}
004	{nationality, dateOfBirth, militaryUnit, militaryGrade}
005	{height, disab_type, weight, selfGrade, eyecolor }
006	{journal, note}
...	...
...	...
030	{nickname, specificDate, disab_percentage, governmentInvolvement, sector}

Fig. 3. Query log converted to sales transaction database.

threshold and the minimum confidence threshold because these values had impact on the results of our experiments. Support reveals the percentage of queries, which contain an attribute set X, i.e., the probability of occurrence of an attribute set X. A frequent attribute set has support, which is greater than or equal to the minimum support threshold, where P denotes the probability.

$$\textbf{Support}(\textbf{X}) = \textbf{P}(\textbf{X})$$

Confidence refers to a conditional probability. This is the probability that the occurrence of an attribute set Y in relationship to the occurrence of an attribute set X. This can be understood as the probability that an attribute set Y occurs in queries given that an attribute set X has already occurred in the same queries. Confidence is used to generate strong association rules from the frequent itemsets that meet the minimum confidence threshold.

$$\textbf{Confidence}(X \Rightarrow Y) = \textbf{support}(X \Rightarrow Y)/\textbf{support}(\textbf{X}) = \textbf{P}(X \cap Y)/$$
$$\textbf{P}(\textbf{X}) = \textbf{P}(Y|X)$$

Like in market basket analysis, false positives and false negatives are inevitable when the thresholds are set either too low or too high, respectively. For example, when the minimum support threshold was too low, false positives tended to occur as those originally infrequent attributes were falsely found as frequent. When the minimum support threshold was too high, false negatives tended to occur where those originally frequent attributes were falsely found as infrequent.

Figure 4 shows the results of one of our experiments, where the minimum support threshold was 12% and the minimum confidence threshold was 55%. In this experiment, 24 strong association rules were generated. Two pairs of attributes were found similar:

- *specificDate* and *sickness_injuryDesc*;
- resource's *title* and resource's *name*.

Among the strong association rules, we looked for sets of attributes that frequently appeared together in the consequents (right sides) of those rules. When we found such rules, we had candidates for similar attributes that appeared in the antecedents (left sides) of those rules. For example, the consequents of Rule 17 and Rule 20 (marked in blue in Fig. 4) suggested a pair of similar attributes: resource's *title* and resource's *name*. Furthermore, the consequents of Rule 11 and Rule 14 (marked in red in Fig. 4) suggested another pair of similar attributes: *specificDate* and *sickness_injuryDesc*. However, intuitively these attributes were not similar and thus, such a finding was a false positive. One possible reason for this false positive was the small size of the query log (which contained only 30 queries).

The size of the query log was a crucial factor that could affect the experiment results. This is because the Apriori algorithm assumes that a sales transaction database is large. It was observed that there was yet another pair of similar attributes but it was not found (false negative): publication's *description* and biography's *description*. Again, due to the small size of the query log, there were no common query patterns, which inquired both attributes.

In summary, using the experiment results, OntoLife could be refined. In particular, OWL has a construct *owl:equivalentProperty* to specify that resource's *title* is a semantic duplicate of resource's *name*.

5 Related Work

5.1 Ahmad El Sayed et al.'s Approach

Ahmad El Sayed et al. [15] proposed to consider the context patterns of text corpus and compute the semantic similarity in taxonomies. The rationale behind their approach is that two concepts, which are similar in one context, may appear dissimilar in another context. The authors distinguished two categories of semantic similarity measure:

- **Knowledge-based measure:** This measure has the following sub-categories:

  - **Edge-based measure:** Here similarities are computed based on the count of edges separating two nodes by an "is-a" relation. However, the measure is not adequate as it considers the taxonomy as a simple structure with uniform edges without exploiting the information.
  - **Node-based measure:** Here similarities are computed based on the similarities of the information associated with each node. This information can be either a node description (so called feature-based measure) or a numerical value augmented from a text corpus (so called information-content measure). Feature-based measures aim to apply on rich resources such as Wordnet because these resources describe most entities and are grouped by synsets.
  - **Information content measure:** Here the similarity between two concepts is computed based on "the extent to which they share common information". The information content value refers to the concept frequency in text corpus.

- **Corpus-based measure:** This measure has the following sub-categories:

Nr.	Association rules:	Confidence:
Support: 13.3%		
1)	$(specificDate \Rightarrow sector)$	100%
2)	$(specificDate \Rightarrow disab\_percentage)$	100%
3)	$(sickness\_injuryDesc \Rightarrow sector)$	100%
4)	$(sickness\_injuryDesc \Rightarrow disab\_percentage)$	100%
5)	$(resource'sname \Rightarrow isAlive)$	80%
6)	$(resource'sname \Rightarrow nickname)$	80%
7)	$(resource'stitle \Rightarrow isAlive)$	57%
8)	$(resource'stitle \Rightarrow nickname)$	57%
9)	$(specificDate, sector \Rightarrow disab\_percentage)$	100%
10)	$(specificDate, disab\_percentage \Rightarrow sector)$	100%
11)	$(specificDate \Rightarrow disab\_percentage, sector)$	100%
12)	$(sickness\_injuryDesc, sector \Rightarrow disab\_percentage)$	100%
13)	$(sickness\_injuryDesc, disab\_percentage \Rightarrow sector)$	100%
14)	$(sickness\_injuryDesc \Rightarrow disab\_percentage, sector)$	100%
15)	$(nickname, resource'sname \Rightarrow isAlive)$	100%
16)	$(isAlive, resource'sname \Rightarrow nickname)$	100%
17)	$(resource'sname \Rightarrow nickname, isAlive)$	80%
18)	$(nickname, resource'stitle \Rightarrow isAlive)$	100%
19)	$(isAlive, resource'stitle \Rightarrow nickname)$	100%
20)	$(resource'stitle \Rightarrow nickname, isAlive)$	57%
Support: 26%		
21)	$(sector \Rightarrow disab\_percentage)$	100%
22)	$(disab\_percentage \Rightarrow sector)$	100%
23)	$(isAlive \Rightarrow nickname)$	88%
24)	$(nickname \Rightarrow isAlive)$	66%

Fig. 4. Experiment results.

- **Co-occurrence-based similarity:** It considers the word co-occurrences in texts.
- **Context-based similarity:** It assumes that similar words will appear in similar contexts. A vector-space model is used as a semantic measuring device.
- **Latent semantic analysis:** It assumes that words that are close in meaning will occur in similar pieces of text. Single value decomposition is used to reduce the number of columns but keeping the similarity structure among words. The similarity of two words is compared by using cosine of the angle between two vectors formed by any two rows.

Since in market-basket analysis one can make two dissimilar items similar by adjusting minimum thresholds, a context can be used to prevent such a problem by ignoring irrelevant information. The authors proposed to adapt semantic similarities to the target corpus as the entities represent the domain of interest in most text-based applications. There are some limitations in an information context measure as it is inaccurate to consider infrequent concepts as more informative than frequent ones. Hence, it is assumed that the taxonomic structure in WordNet is well organized enough to compute an information context measure. There is also a probability of having a concept in a corpus that is not adaptive enough to consider the actual context. Therefore, the context-based approach proposed by Ahmad El Sayed et al. computes semantic similarities by considering the target context from a given text corpus using term frequency and inverse document frequency. It aims at deriving a weighted taxonomy, where the "heavier" subtrees are assumed to have more context representative than the "lighter" ones. Lower similarity value will be given to "heavier" sub-tree than "lighter" ones. In addition, the number of documents where the concept occurred are also being considered. A context-dependency measure is computed for this purpose. Corpus-based word similarities techniques are combined with a context-dependent measure to attain the best possible result. The authors' experiments showed that the results of context-dependency methods can vary from one context to another and need to be refined by human experts.

5.2 Aziz Barbar and Martine Collard's Approach

Aziz Barbar and Martine Collard [16, 17] examined the conversion of a relational database into an object-oriented database and the problem of attribute similarity. They invented a method called UQoRE (User Query oriented Reverse Engineering), which mines a collection of user queries on the database using a data mining algorithm to obtain the distance-similarity measure between two attributes.

Database reverse engineering is a method of producing a new description of stored data by studying the current database implementation. The authors identified various problems of the current database reverse engineering methods, which:

- Do not involve the discovery of the object-oriented conceptual model.
- Make wrong assumptions on the original relational database:

  - Some of the database reverse engineering methods are based on 3NF (Third Normal Form) but real databases are de-normalized for access optimization.
  - Attribute naming is not always consistent.
  - SQL statement analysis can help to obtain database semantics but in a limited extent.
  - Simplification of database reverse engineering methods leads to unwanted non-semantic aspects added to the original aspect, which can cause de-normalized relations.

- Do not consider the structure of entities, the links between entities that can adjust the data to new procedures, fuzzy data. Synonyms/homonyms ambiguities can cause redundancies.

To address the problems above, the authors proposed to extract information on attribute semantics with the help of the queries. This involves the de-optimization of the original database and key extraction. De-optimization involves schema normalization and restructuring. Schema normalization is essential to detect functional dependencies and to spilt the non-3NF relations. As the result of the spilt, new attributes are created, which called for the need of a re-structuring step. This step consists of merging or separating and deleting of some columns. Hence, it is important to identify similar entities using a context-based similarity measure from market basket analysis. In the key extraction, there were several assumptions on equi-join and value-set criteria made. However, it proved to be inadequate in identifying attributes, which either have similar concepts or filtering attributes that have different concepts, e.g., two primary keys linked by a equi-join can have identical value sets and concepts.

The key extraction process involves a specific database, where each record in the database represents a query instance. All user queries are stored in the query database. A 2-dimensional matrix M is used, where a row R of the matrix M represents the queries and a column X of the matrix M represents the attributes found in the queries. A value of 1 at R_j and X_i indicates that an attribute X_i appears in the WHERE clause of a query R_j and a value of 0 indicates otherwise. The similarity distance between two attributes is defined as a measure, which represents the degree of similarity of their context in the queries. That is, two attributes are similar if they are used in similar contexts.

To summarize, Aziz Barbar and Martine Collard's approach is apt to implement old large databases, which require a new design. However, an understanding of the semantic domain of the underlying application by human experts is a prerequisite for this approach.

5.3 Irina Astrova and Arne Koschel's Approach

In our previous work [1, 2], we employed the ICD (Iterated Contextual Distance) algorithm from market basket analysis to detect similar attributes in an ontology. In this algorithm, an arbitrary distance between attributes is initially used to calculate the probability distribution of attributes in a query log. The distribution is re-calculated in subsequent iterations until the algorithm converges. Usually after a few iterations, the algorithm results in a stable distance between the attributes. Such a distance is called iterated contextual distance and ranges between 0 and 1. The value of 0 indicates that attributes are completely similar, whereas the value of 1 indicates that attributes are completely dissimilar. As input, the algorithm takes a query log, which contains M queries with N attributes in the ontology. As output, the algorithm produces a N x N symmetric distance matrix, which contains the iterated contextual distance between the attributes. To detect similar attributes, a threshold is also used. Two attributes are considered as similar if the iterated contextual distance between them is lower than the threshold.

To summarize, the ICD algorithm can detect similar attributes in an ontology. However, the criteria for the convergence of the algorithm are to be examined. On the other hand, a theoretical analysis of the convergence is difficult because the algorithm attempts to compute distances between fixed points of a non-linear dynamic system.

5.4 Luis Cavique's Approach

Luis Cavique [20] invented an algorithm called Similis for market basket analysis. To find frequent itemsets, this algorithm transforms the sales transaction database into a weighted graph, where a vertex set represents an itemset and a weighted edge between vertex sets represents similarity between itemsets.

The Similis algorithm is aimed at solving a drawback of the Apriori algorithm – its exponential time complexity. The Apriori algorithm attempts to reduce the number of itemsets in the input by using the minimum support threshold and the minimum confidence threshold. However, these user-defined parameters depend on the size of sales transaction database and thus, can cause unpredictable data reduction.

The main difference between the Apriori and Similis algorithms is their performance. The latter is faster than the former. To test the performance of the two algorithms, Luis Cavique used the average support of the top five solutions found by the Similis algorithm divided by the average support of the top five solutions found by the Apriori algorithm. The test results showed that the time complexity of the Apriori algorithm increases exponentially as the data volume increases. By contrast, the time complexity of the Similis algorithm is almost constant: it depends only on the size and number of cliques in the weighted graph.

6   Conclusion

During an ontology-based data integration, the newly merged ontology such as OntoLife can contain semantic duplicates (e.g., similar attributes). These attributes can cause incomplete query results. In this paper, we focused on finding semantic duplicates in OntoLife. The approach we proposed is as follows: (1) the Apriori algorithm is applied to find strong association rules derived from a query log, where each query is a set of terms being attributes in the ontology, and (2) candidates for semantic duplicates are determined as antecedents of the strong association rules having similar consequents.

In addition, we conducted a number of experiments on the query log, which was a list of queries posed by users to inquire against OntoLife. On the whole, the results of our experiments showed that market basket analysis in general and the Apriori algorithm in particular are capable of finding similar attributes in the ontology, although user involvement is required to identify the optimal values for the minimum support threshold and the minimum confidence threshold.

Acknowledgement. Irina Astrova's work was supported by the Estonian Ministry of Education and Research institutional research grant IUT33-13.

References

1. Astrova, I., Koschel, A.: Automatic detection of duplicated attributes in ontology. In: Cordeiro, J., Filipe, J. (Eds.) ICEIS 2009: Proceedings of the 11th International Conference on Enterprise Information Systems, Volume DISI. INSTICC, 2009, pp. 283–286 (2009)
2. Astrova, I.: Improving query results with automatic duplicate detection. In: Ioannidis, Y., Manghi, P., Pagano, P. (Eds.) Proceedings of the Second Workshop on Very Large Digital Libraries, VLDL 2009: A Workshop in conjunction with the European Conference on Digital Libraries 2009. Institute of Information Science and Technology; DELOS Association (2009)

3. Agrawal, R., Srikant, R.: Fast Algorithms for Mining Association Rules In: Stonebraker, M., Hellerstein, J.M. (Eds.) Readings in database systems (3rd ed.). Morgan Kaufmann Publishers Inc., San Francisco, CA, USA, pp. 580–592 (1998)
4. Kargiot, E., Kontopoulos, E.: OntoLife: an ontology for semantically managing personal information. http://lpis.csd.auth.gr/ontologies/ontolist.html#ontolife
5. Person Ontology. http://ebiquity.umbc.edu/ontology/person.owl
6. Friend of a Friend (FOAF) Ontology. http://xmlns.com/foaf/spec/
7. Family Tree Ontology. http://users.auth.gr/elkar/thesis/FamilyTree.owl
8. Relationship Ontology. http://purl.org/vocab/relationship/
9. ISO lists for Countries and Languages Ontology. http://psi.oasis-open.org/iso/639/#
10. Project Ontology. http://ebiquity.umbc.edu/ontology/project.owl
11. Research Ontology. http://ebiquity.umbc.edu/ontology/research.owl
12. Publication Ontology. http://ebiquity.umbc.edu/ontology/publication.owl
13. PersonProjectAssociation Ontology. http://ebiquity.umbc.edu/ontology/association.owl
14. Biography Ontology. http://users.auth.gr/elkar/thesis/Biography.owl
15. El Sayed, A., et al.: A new context-aware measure for semantic distance using a taxonomy and a text corpus. In: Proceedings of IRI, pp. 279–284 (2007)
16. Barbar, A., Collard, M.: A distance-based approach for database re-engineering. In: Proceedings of the ACS/IEEE International Conference on Computer Systems and Applications (AICCSA 2001). IEEE Computer Society, Washington, DC, USA, pp. 188–190 (2001)
17. Barbar, A., Collard, M.: Semantic extraction: a user-driven method (2001). http://www.fit.vutbr.cz/events/ism/2001/pdf/barbar.pdf
18. Khan, Z.C., Keet, C.M.: SUGOI: automated ontology interchangeability. In: Knowledge Engineering and Knowledge Management, pp. 150–153 (2015)
19. Mascardi, V., Locoro, A., Rosso, P.: Automatic ontology matching via upper ontologies: a systematic evaluation. IEEE Trans. Knowl. Data Eng. **22**(5), 609–623 (2010)
20. Cavique, L.: Graph-based structures for the market baskets analysis (2004). http://lcavique.no.sapo.pt/publicacoes/Similis%20APDIO.pdf

Correction to: Deploying Personalized Oblivious Mobile Identity

Georgia Sourla and Evangelos Sakkopoulos ⓘD

Correction to:
Chapter "Deploying Personalized Oblivious Mobile Identity"
in: M. Virvou et al. (Eds.): *Knowledge-Based Software*
***Engineering: 2020*, LAIS 19,**
https://doi.org/10.1007/978-3-030-53949-8_15

In the original version of the book, the following belated correction has been incorporated:

In Chapter 15, the affiliation of author "Georgia Sourla" has been changed from "Scytales, Argyroupoli, Sweden" to "Scytales AB, Taby, Sweden".
The chapter and book have now been updated with the change.

The updated version of this chapter can be found at
https://doi.org/10.1007/978-3-030-53949-8_15

Author Index

© Springer Nature Switzerland AG 2020
M. Virvou et al. (Eds.): JCKBSE 2020, LAIS 19, pp. 193–194, 2020.
https://doi.org/10.1007/978-3-030-53949-8

Printed in the United States
by Baker & Taylor Publisher Services